Innovation, Technology, and Knowledge Management

Series Editor

Elias G. Carayannis
School of Business
George Washington University
Washington, DC, USA

More information about this series at http://www.springer.com/series/8124

Marta Peris-Ortiz • João J. Ferreira
Jose M. Merigó Lindahl
Editors

Knowledge, Innovation and Sustainable Development in Organizations

A Dynamic Capabilities Perspective

Springer

Editors
Marta Peris-Ortiz
Department of Business Administration
Universitat Politècnica de València
Valencia, Spain

João J. Ferreira
Management and Economics Department
Universidade da Beira Interior
Covilhã, Portugal

Jose M. Merigó Lindahl
Department of Management Control
and Information Systems
University of Chile
Santiago, Chile

ISSN 2197-5698 ISSN 2197-5701 (electronic)
Innovation, Technology, and Knowledge Management
ISBN 978-3-030-09095-1 ISBN 978-3-319-74881-8 (eBook)
https://doi.org/10.1007/978-3-319-74881-8

Printed on acid-free paper

This Springer imprint is published by the registered company Springer International Publishing AG part of Springer Nature.
The registered company address is: Gewerbestrasse 11, 6330 Cham, Switzerland

Series Foreword

The Springer book series *Innovation, Technology, and Knowledge Management* was launched in March 2008 as a forum and intellectual, scholarly "podium" for global/local, transdisciplinary, transsectoral, public–private, and leading/"bleeding"-edge ideas, theories, and perspectives on these topics.

The book series is accompanied by the Springer *Journal of the Knowledge Economy*, which was launched in 2009 with the same editorial leadership.

The series showcases provocative views that diverge from the current "conventional wisdom," that are properly grounded in theory and practice, and that consider the concepts of ***robust competitiveness***,[1] ***sustainable entrepreneurship***,[2] and ***democratic capitalism***,[3] central to its philosophy and objectives. More specifically, the aim of this series is to highlight emerging research and practice at the dynamic intersection of these fields, where individuals, organizations, industries, regions, and nations are harnessing creativity and invention to achieve and sustain growth.

---

[1] We define *sustainable entrepreneurship* as the creation of viable, profitable, and scalable firms. Such firms engender the formation of self-replicating and mutually enhancing innovation networks and knowledge clusters (innovation ecosystems), leading toward robust competitiveness (E.G. Carayannis, *International Journal of Innovation and Regional Development* 1(3), 235–254, 2009).

[2] We understand *robust competitiveness* to be a state of economic being and becoming that avails systematic and defensible "unfair advantages" to the entities that are part of the economy. Such competitiveness is built on mutually complementary and reinforcing low-, medium-, and high-technology and public and private sector entities (government agencies, private firms, universities, and nongovernmental organizations) (E.G. Carayannis, *International Journal of Innovation and Regional Development* 1(3). 235–254, 2009).

[3] The concepts of *robust competitiveness* and *sustainable entrepreneurship* are pillars of a regime that we call *democratic capitalism* (as opposed to "popular or casino capitalism"), in which real opportunities for education and economic prosperity are available to all, especially – but not only – younger people. These are the direct derivative of a collection of top-down policies as well as bottom-up initiatives (including strong research and development policies and funding, but going beyond these to include the development of innovation networks and knowledge clusters across regions and sectors) (E.G. Carayannis and A. Kaloudis, *Japan Economic Currents*, p. 6–10 January 2009).

Books that are part of the series explore the impact of innovation at the "macro" (economies, markets), "meso" (industries, firms), and "micro" levels (teams, individuals), drawing from such related disciplines as finance, organizational psychology, research and development, science policy, information systems, and strategy, with the underlying theme that for innovation to be useful it must involve the sharing and application of knowledge.

Some of the key anchoring concepts of the series are outlined in the figure below and the definitions that follow (all definitions are from E.G. Carayannis and D.F.J. Campbell, *International Journal of Technology Management*, 46, 3–4, 2009).

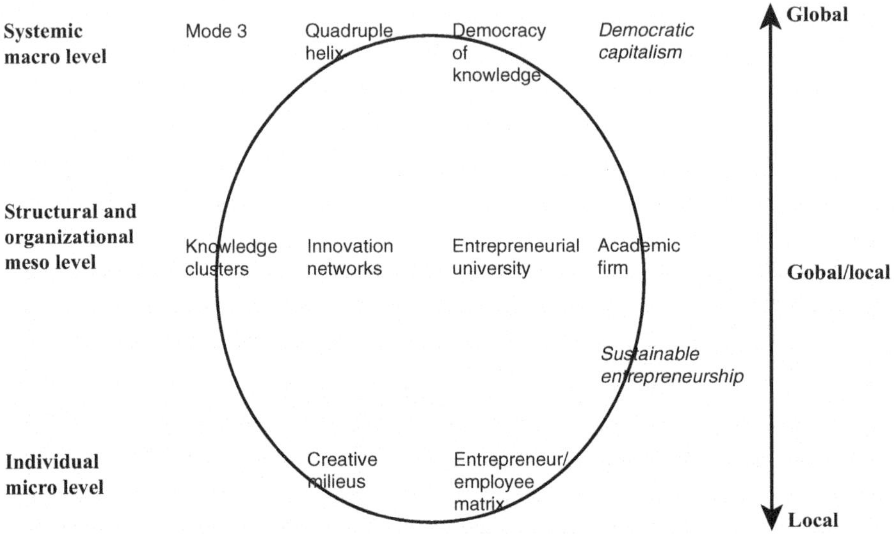

Conceptual profile of the series *Innovation, Technology, and Knowledge Management*

- The "Mode 3" Systems Approach for Knowledge Creation, Diffusion, and Use: "Mode 3" is a multilateral, multinodal, multimodal, and multilevel systems approach to the conceptualization, design, and management of real and virtual, "knowledge-stock" and "knowledge-flow," modalities that catalyze, accelerate, and support the creation, diffusion, sharing, absorption, and use of cospecialized knowledge assets. "Mode 3" is based on a system-theoretic perspective of socioeconomic, political, technological, and cultural trends and conditions that shape the coevolution of knowledge with the "knowledge-based and knowledge-driven, global/local economy and society."
- Quadruple Helix: Quadruple helix, in this context, means to add to the triple helix of government, university, and industry a "fourth helix" that we identify as the "media-based and culture-based public." This fourth helix associates with "media," "creative industries," "culture," "values," "life styles," "art," and perhaps also the notion of the "creative class."

- Innovation Networks: Innovation networks are real and virtual infrastructures and infratechnologies that serve to nurture creativity, trigger invention, and catalyze innovation in a public and/or private domain context (for instance, government–university–industry public–private research and technology development coopetitive partnerships).
- Knowledge Clusters: Knowledge clusters are agglomerations of cospecialized, mutually complementary, and reinforcing knowledge assets in the form of "knowledge stocks" and "knowledge flows" that exhibit self-organizing, learning-driven, dynamically adaptive competences and trends in the context of an open systems perspective.
- Twenty-First Century Innovation Ecosystem: A twenty-first century innovation ecosystem is a multilevel, multimodal, multinodal, and multiagent system of systems. The constituent systems consist of innovation metanetworks (networks of innovation networks and knowledge clusters) and knowledge metaclusters (clusters of innovation networks and knowledge clusters) as building blocks and organized in a self-referential or chaotic fractal knowledge and innovation architecture (Carayannis 2001), which in turn constitute agglomerations of human, social, intellectual, and financial capital stocks and flows as well as cultural and technological artifacts and modalities, continually coevolving, cospecializing, and cooperating. These innovation networks and knowledge clusters also form, reform, and dissolve within diverse institutional, political, technological, and socioeconomic domains, including government, university, industry, and nongovernmental organizations and involving information and communication technologies, biotechnologies, advanced materials, nanotechnologies, and next-Generation energy technologies.

Who is this book series published for? The book series addresses a diversity of audiences in different settings:

1. *Academic communities:* Academic communities worldwide represent a core group of readers. This follows from the theoretical/conceptual interest of the book series to influence academic discourses in the fields of knowledge, also carried by the claim of a certain saturation of academia with the current concepts and the postulate of a window of opportunity for new or at least additional concepts. Thus, it represents a key challenge for the series to exercise a certain impact on discourses in academia. In principle, all academic communities that are interested in knowledge (knowledge and innovation) could be tackled by the book series. The interdisciplinary (transdisciplinary) nature of the book series underscores that the scope of the book series is not limited a priori to a specific basket of disciplines. From a radical viewpoint, one could create the hypothesis that there is no discipline where knowledge is of no importance.
2. *Decision makers – private/academic entrepreneurs and public (governmental, subgovernmental) actors:* Two different groups of decision makers are being addressed simultaneously: (1) private entrepreneurs (firms, commercial firms, academic firms) and academic entrepreneurs (universities), interested

in optimizing knowledge management and in developing heterogeneously composed knowledge-based research networks; and (2) public (governmental, subgovernmental) actors that are interested in optimizing and further developing their policies and policy strategies that target knowledge and innovation. One purpose of public *knowledge and innovation policy* is to enhance the performance and competitiveness of advanced economies.

3. *Decision makers in general:* Decision makers are systematically being supplied with crucial information, for how to optimize knowledge-referring and knowledge-enhancing decision-making. The nature of this "crucial information" is conceptual as well as empirical (case-study-based). Empirical information highlights practical examples and points toward practical solutions (perhaps remedies), conceptual information offers the advantage of further-driving and further-carrying tools of understanding. Different groups of addressed decision makers could be decision makers in private firms and multinational corporations, responsible for the knowledge portfolio of companies; knowledge and knowledge management consultants; globalization experts, focusing on the internationalization of research and development, science and technology, and innovation; experts in university/business research networks; and political scientists, economists, and business professionals.
4. *Interested global readership:* Finally, the Springer book series addresses a whole global readership, composed of members who are generally interested in knowledge and innovation. The global readership could partially coincide with the communities as described above ("academic communities," "decision makers"), but could also refer to other constituencies and groups.

Washington, DC, USA Elias G. Carayannis

Contents

Chapter 1
Knowledge, Innovation, and Sustainable Development in Organizations: A Dynamic Capability Perspective: An Overview

Marta Peris-Ortiz, João J. Ferreira, and Jose M. Merigó Lindahl

1.1 Introduction

The chapters compiled in this book discuss the change which has taken place in knowledge and innovation in the core of different companies and how this change alters the companies themselves and the markets in which they compete from a dynamic capability perspective.

At the same time, it takes into account how these changes must preserve the conditions of the ecosystems which make sustainability possible (Kiron et al. 2013; Kovel 2007). The necessities of sustainability initially appear as a restriction; however in the degree in which the legislation changes to preserve the environment and the citizens' awareness increases, modifying their consumer behaviors, these new characteristics of the environment and the markets can be transformed into a new set of opportunities. This means that the classic academic contributions about new combinations of factors (Schumpeter 1934) or about how management absorbs the experiences associated with the new environments (Penrose 1959) must be reconsidered in the light of the preservation of the technological, ecological, and social ecosystems.

M. Peris-Ortiz (✉)
Department of Business Administration, Universitat Politècnica de València, Valencia, Spain
e-mail: mperis@doe.upv.es

J. J. Ferreira
Management and Economics Department, University of Beira Interior, Covilhã, Portugal
e-mail: jjmf@ubi.pt

J. M. Merigó Lindahl
Department of Management Control and Information Systems, University of Chile, Santiago, Chile
e-mail: jmerigo@fen.uchile.cl

M. Peris-Ortiz et al. (eds.), *Knowledge, Innovation and Sustainable Development in Organizations*, Innovation, Technology, and Knowledge Management,
https://doi.org/10.1007/978-3-319-74881-8_1

The resource-based theory (RBV) foundation is that companies are heterogeneous regarding their resources and capabilities, which are the basis of the firm's growth and competitive advantage. Therefore in recent years, the strategy literature has placed considerable interest to understand how companies create, maintain, and enhance these capabilities. And it is also fitting to say how companies shape their competitive advantage from organizational learning (Nonaka 1994; Crossan et al. 1999), from the idiosyncratic mixture of their capabilities (Peteraf 1993; Barney 2001), or from their dynamic capabilities (Teece 2007). In all cases, the organizational learning, the idiosyncratic mixture which preserves the competitive advantage, and the dynamic capabilities are new forms of knowledge and innovation when they achieve their objective and lead to practical actions; and they must also be reconsidered in the light of sustainability, because the incremental or radical innovations which take place in the companies affect both the processes and the products as well as their incidence in the environment. All the forms of learning – individual, group, or organizational – need to be modified to create forms of knowledge and sustainable practices.

Finally, it is fitting to add another major source to the sources of innovation which act from the internal dimension of the company, this time, an external source which we know as open innovation (Cheng and Huizingh 2014) and which has a wide variety of forms: ranging from those which consist in incorporating new knowledge and technologies without interaction or with a slightly relevant interaction with the external sources up to those which require a strong and prolonged relation (Dahlander and Gann 2010). In all these cases as in those which refer to the internal sources of innovation, the relations and forms of open innovation will be modified by the requirements of sustainability (Kovel 2007).

The discussed questions are some of the issues which have a major relation with the processes and the dynamics of knowledge and innovation and how this supposes a major source of change, both in the acquisition of competitive advantages for the company and the transformation of the ecological and social environment; hence the different chapters of this book aim to compile different processes and relevant forms for the creation of knowledge and innovation and their relation with sustainability. In order for companies to obtain competitive advantages which are their fundamental purpose, at present, it is not sufficient to innovate; jointly with innovation, companies must convert the preservation of their business ecosystem into another essential source of competitive advantage. Nothing inaugurates as many opportunities as changing the world so that it is sustainable.

The book aims to collect the most recent research and best practices in the cooperative and networking small business field identifying new theoretical models and describing the relationship between cooperation and networks in small business strategy context.

1.2 Theoretical Background

The existence of a solid strategy is essential to develop a business value, thus highlighting its main competitors. A business strategy requires knowledge about the company's resources and its competencies; thus each resource contributes to the formation of forces resulting in the development of a sustainable competitive advantage (Ducan et al. 1998). Penrose (1959) advocates the need for a company to develop internal resources as exploratory tools of knowledge and innovation, so that it can add strategic value and positioning.

Little importance was given to knowledge, as an intangible asset, until the 1980s; however, today it is classified as a valuable and irreplaceable resource for the continued growth of organizations that increasingly act in uncertain and unpredictable contexts where strategic information management is rewarded, such as innovation and ownership of knowledge. Clearly, in a world where markets, products, technology, and society itself are rapidly changing, knowledge has become the main source of sustainable competitive advantage (Nonaka and Teece 2001; Vorakulpipat and Rezgui 2008).

Therefore, with the manifestation of the age of knowledge, everyone has recognized that the intangible assets of an organization will be essential in both the capability to create competitive advantages and growth at a rapid pace. As a consequence, organizations demonstrate increasingly more special attention to value creation, through the power of knowledge (Ambrosini et al. 2009; Schiuma and Lerro 2008). Since this is an important source of sustainable competitive advantage, it is necessary to develop ways of identifying, creating, evaluating, and applying this resource, as well as designing appropriate systems for its management.

Knowledge management, as a management attitude, forms an essential part of the process in the organizational strategy that determines the management of people and information and communication technologies (Davenport and Prusak 1998; Soliman and Spooner 2000), with a view to promoting integrated organizational learning.

Therefore, knowledge transfer is fundamental to the performance of knowledge creation and knowledge capture for better organizational performance (Von Krogh et al. 2000), and the effectiveness of organizational knowledge transfer is influenced by key organizational factors, such as structure, culture, processes and strategy, and information technologies (Ives et al. 2003).

As markets change, technologies multiply, competitors increase, and products fall into disuse almost overnight. Successful organizations are those that continually create new knowledge, which disseminate it unrestricted throughout the organization and quickly apply it to new technologies and products (Nonaka 1991). It is these activities, creation, dissemination, and insertion of knowledge, that define an organization as a creator of knowledge whose exclusive business is continuous innovation (Nonaka 1991).At present, in the competitive landscape in which we live, innovation becomes essential. A superior ability to innovate provides organizations with opportunities to grow faster, better, and smarter than their competitors (Davila et al. 2006; Sáenz et al. 2009).

Since the works of Nonaka (1991), the concept of innovation has been closely related to that of knowledge creation. Along this line, it is generally assumed that the innovation process consists of a permanent exercise of new and unique knowledge (Sáenz et al. 2009). In particular, knowledge creation involves a continuous process, whereby a person goes beyond the constraints and individual boundaries imposed by information and past learning through the acquisition of new contexts, a new worldview, and new knowledge.

The examination for sources of competitive advantage has been the object of many studies, and one of the most recent approaches focuses on the paradox of resources and capabilities (Barney 1991; Ferreira and Fernandes 2017). However, whether in more or less static environments, the identification of competitive advantage can become relatively easy to identify; in dynamic environments the organizational ability to sustain and renew competitive advantage has become much more complex but is imperative to formulate it.

Resource-based view (RBV) thus analyzes a company from the point of view of its internal resources, considered fundamental factors for innovation. The RBV had its origin in Penrose (1959), which introduced a new concept of enterprise, as a set of resources available to an administrative unit.

Teece and Pisano (1994) argue that the RBV does not provide explanations on how some companies can gain competitive advantage in situations of rapid and turbulent changes and, from this perspective, the perspective of the dynamic capabilities that seeks to explain the strategic position and origin of the advantage competitive (Eisenhardt & Martin 2000; Teece et al. 1997). According to Teece and Pisano (1994), dynamic capabilities are defined as the organization's ability to integrate, build, and reconfigure internal and external competencies to respond rapidly to changing environments.

Likewise, Eisenhardt and Martin (2000) define them as the organizational processes that use resources, specifically to integrate, reconfigure, obtain, or release resources that adapt or even create changes in the market. Barney (1991) argues that firms that own and exploit valuable and rare capabilities will achieve competitive advantage by concluding that these advantages manifest themselves in better short-term performance. In this sense, RBV is considered to be one of the most accepted strategic management theories presenting resources and capabilities as an essential source for obtaining a sustainable competitive advantage and, consequently, a higher performance for companies.

According to Helfat and Peteraf (2003), capabilities can go through a cycle of events, where the first stage is characterized by capacity building, then development, and ultimately capacity maturity. This approach was developed to explain the need for companies to continue to develop and improve their skills, trying to defend themselves against competitors' attempts to duplicate the capabilities that generate competitive advantage (Collis 1994).

Resources and capabilities are long-lasting, implying that a resource-based strategy provides a long-term vision compared to the traditional perspective. This new perspective becomes more robust in dynamic and uncertain environments, making it more sustainable. Recently, Zahra et al. (2006) define dynamic capabilities as the ability to reconfigure the resources and routines of the company in a manner considered and appropriate by its main decision maker. These authors place managers at the center of this process by assigning them a crucial role in the choices and willingness to productively transform existing routines or resources and implement them. According to Zahra and George (2002), absorption capacity provides a competitive advantage to the company in terms of strategic flexibility, innovation, and performance.

Vega-Jurado et al. (2008) reinforce the discussion that absorption capacity is determined not only by R&D activities but also including the applicability of the available knowledge in the environment as a determining factor. These authors point to the relationship between absorption capacity and the complexity of knowledge, that is, complex knowledge provokes stronger interactions, whereas for more elementary knowledge, this type of interaction would not be required.

Absorption capacity promotes the continual renewal of the inventory of critical knowledge for the organization's future activities and allows the visualization of new strategic choices (Jansen et al. 2005). Thus, the greater the company's ability to interpret and understand the information obtained from external sources, the greater its chances of anticipating changes, thus broadening its response options to perceived variations in the environment (Ben-Menahem et al. 2013).

1.3 Overview of Book Contents

This book includes 13 chapters relating to the knowledge, innovation, and sustainable development in a dynamic capability perspective that covers a broad range of organizational contexts. Furthermore, the chapters in this book examine the theme using different theoretical backgrounds and different methodologies. Individually, each chapter gives rich insights about the phenomenon they explore.

This current chapter (Chap. 1) attempts to show an overview about this topic, and it address the main critical aspects approached in the different chapters and their impact around these themes.

Chapter 2 undertaken by Cristina Fernandes, João J. Ferreira, and Marta Peris-Ortiz entitled "Knowledge, Innovation, and Sustainability: Past Literature and Future Trends" analyzes through a bibliometric perspective the scientific field of this topic in order to better grasp how these concepts and dimensions have evolved over time. The contribution of this chapter is precisely to find the theoretical approaches present in the different investigations on this field and to promote the discussion of future agenda.

Chapter 3, entitled "Project Management Office in the Nongovernmental Organization as a Driver of Sustainable Competitive Advantage: A Dynamic Capabilities Approach" by Adonai J. Lacruz, Everton A. Cunha, Ralf L. de Moura, and Marcos P. V. de Oliveira, studies the impact of project management office

(PMO) on the performance of projects of nongovernmental organizations (NGOs), under the lens of dynamic capabilities. The results lead to the proposition that the PMOs in NGOs act as catalysts of dynamic capacities, allowing these organizations to reach higher levels of performance in their projects. The authors argue that the PMO has contributed to making the NGO able to perceive opportunities and threats, avail opportunities, and maintain the conditions of competitiveness by the continuous improvement of the business assets, leading to the obtaining sustainable competitive advantage.

Chapter 4, carried out by Mário Franco and Heiko Haase, entitled "Sustainable Development of Small and Medium-Sized Enterprises in Disadvantaged Regions: Impact of Knowledge and Innovation," focuses on a qualitative approach through multiple small and medium-sized enterprise (SME) cases; the authors attempt to analyze how knowledge and innovation of SME can stimulate the growth and development of less favored regions. They found that focusing on knowledge and innovation can be a route to SME survival and success, as this allows the exploitation of new markets, new products, equipment, or processes, maintaining competitiveness at the national level and giving access to international markets. Furthermore, SME strategies rely significantly on the capability to manage knowledge, entrepreneurship, and innovation as a way to improve the quality of their products and provide a broader supply.

Chapter 5 by Jordi Mauri-Castello, Antonio Alonso-Gonzalez, and Marta Peris-Ortiz, entitled "Applied Innovation Methodology: A Proposal for a Dynamic Sustainable Environment for the Generation of Innovation and Knowledge Management Practices in SMEs," aims to describe how a small and medium-sized enterprise could complement the knowledge and dynamic capability management of the customer with a strategic point of view and the ability to overcome any inconvenient challenge or problem. To this end, a new model called applied innovation methodology was proposed, in order to create dynamic and sustainable new opportunities and to enhance the best practices of innovation and knowledge management within these types of organizations.

Chapter 6, entitled "Following the Footprints of SME Competitiveness in a High-Technology Sector" by Luis Farinha and Sharmistha Bagchi-Sen, proposes through a qualitative and quantitative approach to analyze the innovative and entrepreneurial collaborative trends associated with a competitive high-tech cluster that operates in the international market. The authors evidence the importance of collective efficiency strategies, with a focus on RD&I, from a universe of SMEs. Furthermore, lessons about the role of innovation in the SME competitiveness and the role of innovation and entrepreneurship networks in the enhancing of a high-technology industrial cluster were suggested by the authors.

Chapter 7 by Maria D. Moreno-Luzon, Juan P. Escorcia-Caballero, and Odette Chams-Anturi, entitled "The Integration of the Supply Chain as a Dynamic Capability for Sustainability: The Case of an Innovative Organic Company," aims to understand the role of supplier integration in the implementation of radical innovation. For this purpose, the authors first propose a theoretical model to connect supplier integration as a dynamic capability to ordinary routines and, secondly, study the connection between supplier integration and radical innovation through a case study in an innovative firm in the organic agro-food industry. They conclude that it is important to study supplier integration as a dynamic capability from a bidirectional viewpoint, considering the ability of the company to assimilate external knowledge and adapt its resources but, at the same time, understanding how buyers provide knowledge and help their suppliers so they can adapt their resources to cater for new requirements.

Chapter 8, entitled "Personal Branding as a Knowledge Management Tool to Enhance Innovation and Sustainable Development in Organizations," by Antonio Alonso-Gonzalez, Sofía Estelles-Miguel, and José J. Cao-Alvira, proposes some considerations to be taken into account to take advantage of personal branding and its benefits in the context of an organization. The authors argue that personal branding should be implemented by the companies' human resources department through a proper and efficient knowledge management and dynamic capability implementation, in order to promote an acceptable work environment that fosters innovation, sustainable development, transparency, competitiveness, initiative, teamwork, and corporate value alignment within the organization.

Chapter 9, by Silveli C. Andrade, Bruno Felix, and Emerson W. Mainardes, entitled "Expatriation Knowledge Management: The Role of Openness to Change and Work Engagement," proposes a theoretical model that connects the openness to change personality trait as the antecedent to the intention of an individual to be expatriated or also an antecedent to leave the company. This addresses relevant aspect for knowledge management practices in international organizations. The authors argue that the personality trait openness to change offers a paradoxical feature in expatriation candidates in an organization. They also suggest that professionals with a high degree of openness to change tend to be simultaneously more likely to be expatriates as well as presenting a high propensity to leave the company.

Chapter 10, entitled "The Entrepreneurial University Stimulating Innovation Through Campus Development: The MIT Case," by Flavia T. J Curvelo Magdaniel, examines the university campus as a strategic organizational resource helping universities to adapt from a traditional to an entrepreneurial profile in order to remain competitive in today's knowledge-based economy. This is linked to the concept of dynamic capabilities with corporate real estate management theories to study campus development as a long-term process enabling universities to adapt to the changing environments in which they operate. The findings show how the Massachusetts Institute of Technology (MIT) developed its campus and surroundings into a rich innovation ecosystem by reacting timely and adequately to technological, societal, and environmental dynamics influencing its local context.

Chapter 11, undertaken by Victoria Kopeikina, Paula O. Fernandes, and Olga Kosenchuk, entitled "Increasing the Economic Sustainability of the Company JSC Novoazovskoe," explores the meaning of "economic sustainability" and how to enhance its growth taking into account the modification of the harvesting system. It has proposed the modern system of haylage harvesting, offered the purchase of required agricultural machineries to implement the new system, and presented the economic results received after implementation of the new haylage harvesting system.

Chapter 12, entitled "Innovation Management in Portuguese and Russian Agricultural Companies," written by Mikhail Kopeykin, Sofia Cardim, Vitaly Aleshchenko, and Frederico Branco, analyzed the opportunity of implementation of renewable energy sources on existing agricultural enterprises. A comparison between two countries was made. The authors aim to searching of theoretical statements and validation measures of innovative activity increasing in Russian and Portuguese enterprises by implementing the renewable source of energy, namely, solar panels. They found that one of the main features of innovative activity is the opportunity to implement the most recent technologies and equipment and to use the most recent information.

The last chapter (Chap. 13), entitled "Sustainability and Innovation in the Value Chain: An Analysis of a Case Study," by Ronnie J-Figueiredo, Osvaldo Quelhas, and Bouchaib Bahli, presents an innovative practice for creating organizational knowledge and achieving sustainable competitive advantage. The authors address the theoretical relationship between circular economy, strategic innovation, sustainability, competitiveness, and dynamic capabilities, through a case study of a global and innovative company in the energy sector in Brazil.

1.4 Conclusions

Knowledge, innovation, and sustainable development are constructs intricately linked to each other, and the analysis of issues at their interface is crucial to understand the best practices of dynamic capabilities in an organizational context. This book offers a wide-ranging, well-organized, and richly illustrated study of knowledge, innovation, and sustainability of organizations in a perspective of dynamic capabilities.

Taking into account that organizations are inserted in environmental complexity and uncertainty, the capability to adapt to these contingencies is essential. The greater their adaptability, the greater their competitive advantage over their competitors. One way to achieve this competitive advantage is precisely through knowledge and innovation. Organizations must understand that in order to be in the vanguard of knowledge, they have to go and look for the sources where that knowledge is actually generated, which means, in the academia. Moreover, when organizations increase their innovative capability, they will contribute to a better performance and, consequently, enhanced competitive positioning. Accordingly, when entrepreneurs become aware of the importance of knowledge, innovation, and sustainability in a dynamic capability perspective, they will have more opportunities to achieve this highly sought differentiation and, consequently, a sustainable competitive advantage.

We expect this book to provide an important contribution to academic researchers and policy makers by offering a comprehensive understanding of how and why knowledge, innovation, and sustainable development in organizations in a dynamic capability perspective have a crucial meaning for the very foundations of the small business, management, and economics field.

References

Ambrosini, V., Collier, N., & Jenkins, M. (2009). A configurational approach to the dynamics of firm level knowledge. *Journal of Strategy and Management, 2*(1), 4–30.

Barney, J. (1991). Firm resources and sustained competitive advantage. *Journal of Management, 17*, 99–120.

Barney, J. (2001). Is the resource-based "view" a useful perspective for strategic management research? Yes. *Academy of Management Review, 26*(1), 41–56.

Ben-Menahem, S. M., Kwee, Z., Volberda, H. W., & Van Den Bosch, F. A. (2013). Strategic renewal over time: The enabling role of potential absorptive capacity in aligning internal and external rates of change. *Long Range Planning, 46*(3), 216–235.

Cheng, C. C. J., & Huizingh, E. K. R. E. (2014). When is open innovation beneficial? The role of strategic orientation. *Journal of Product Innovation Management, 31*(6), 1235–1253.

Collis, D. (1994). Research note: How valuable are organizational capabilities? *Strategic Management Journal, 15*, 143–152.

Crossan, M. M., Lane, H. W., & White, R. E. (1999). An organizational learning framework: From intuition to institution. *Academy of Management Review, 24*(3), 522–537.

Dahlander, L., & Gann, D. M. (2010). How open is innovation? *Research Policy, 39*, 699–709.

Davenport, T., & Prusak, L. (1998). *Working knowledge: How organizations manage what they know*. Boston, MA: Harvard Business School Press.

Davila, T., Epstein, M., & Shelton, R. (2006). *Making innovation work: How to manage it, measure it, and profit from it*. Upper Saddle River, NJ: Pearson Education.

Duncan, W. J., Ginter, P. M., & Swayne, L. E. (1998). Competitive advantage and internal organizational assessment. *Academy of Management Executive, 12*(3), 6–16.

Eisenhardt, K. M., & Martin, J. (2000). Dynamic capabilities: What are they? *Strategic Management Journal, 21*(10–11), 1105–1121.

Ferreira, J., & Fernandes, C. (2017). Resources and capabilities' effects on firm performance: What are they. *Journal of Knowledge Management, 21*(5), 1202–1217.

Helfat, C., & Peteraf, M. (2003). The dynamic resource-based view: Capability lifecycles, *Strategic Management Journal*, October Special, *24*, 997–1010.

Ives, W., Torrey, B., & Gordon, C. (2003). Knowledge transfer: Transfer is human behavior. In C. Morey, M. Maybury, & B. Thuraisingham (Eds.), *Knowledge management: Classic and contemporary works*. Cambridge, MA: MIT Press.

Jansen, J. J. P., Van Den Bosch, F. A. J., & Volberda, H. W. (2005). Managing potential and realized absorptive capacity: How do organizational antecedents matter. *Academy of Management Journal, 48*(6), 999–1015.

Kiron, D., Kruschwitz, N., Reeves, M., & Goh, E. (2013). The benefits of sustainability-driven innovation. *MIT Sloan Management Review, 54*(2), 69–73.

Kovel, J. (2007). *The enemy of nature: The end of capitalism or the end of the world?* New York: Zed Books Ltd.

Nonaka, I. (1991). The knowledge creating company. *Harvard Business Review, 69*(6), 96–104.

Nonaka, I. (1994). A dynamic theory of organizational knowledge creation. *Organization Science, 5*(1), 14–37.

Nonaka, I., & Teece, D. (2001). Research directions of knowledge management. In I. Nonaka & D. Teece (Eds.), *Managing industrial knowledge: Creation, transfer and utilization* (pp. 330–335). London: Sage.

Penrose, E. (1959). *The theory of the growth of the firm*. New York: Oxford University Press.

Peteraf, M. A. (1993). The cornerstone of competitive advantage: A resource-based view. *Strategic Management Journal, 14*(3), 179–191.

Sáenz, J., Aramburu, N., & Rivera, O. (2009). Knowledge sharing and innovation performance: A comparison between high-tech and low-tech companies. *Journal of Intellectual Capital, 10*(1), 22–36.

Schiuma, G., & Lerro, A. (2008). Knowledge-based capital in building regional innovation capacity. *Journal of Knowledge Management, 12*(5), 121–136.
Schumpeter, J. (1934). *The theory of economic development*. Cambridge, MA: Harvard University Press.
Soliman, F., & Spooner, K. (2000). Strategies for implementing knowledge management: Role of human resources management. *Journal of Knowledge Management, 4*(4), 337–345.
Teece, D. J. (2007). Explicating dynamic capabilities: The nature and microfoundation of (sustainable) enterprise performance. *Strategic Management Journal, 28*(13), 1319–1350.
Teece, D. J., & Pisano, G. (1994). The dynamic capabilities of firms: An introduction. *Industrial and Corporate Change, 3*(3), 537–556.
Teece, D. J., Pisano, G., & Shuen, A. (1997). Dynamic capabilities and strategic management. *Strategic Management Journal, 18*(7), 509–533.
Vega-Jurado, J., Gutiérrez-Gracia, A., & Fernándes-De-Lucio, I. (2008). Analyzing the determinants of firm's absorptive capacity: Beyond R&D. *R&D Management, 38*(4), 392–405.
Von Krogh, G., Nonaka, I., & Ichijo, K. (2000). *Enabling knowledge creation: How to unlock the mystery of tacit knowledge and release the power of innovation*. New York: Oxford University Press.
Vorakulpipat, C., & Rezgui, Y. (2008). An evolutionary and interpretive perspective to knowledge management. *Journal of Knowledge Management, 12*(3), 17–34.
Zahra, S. A., & George, G. (2002). Absorptive capacity: A review, reconceptualization, and extension. *Academy of Management Review, 27*(2), 185–203.
Zahra, S. A., Sapienza, H. J., & Davidsson, P. (2006). Entrepreneurship and dynamic capabilities: A review, model and research agenda. *Journal of Management Studies, 43*(4), 917–955.

Chapter 2
Knowledge, Innovation and Sustainability: Past Literature and Future Trends

Cristina Fernandes, João J. Ferreira, Pedro M. Veiga, and Marta Peris-Ortiz

2.1 Introduction

The interest in studying the importance of knowledge recently emerged as a new subject field in its own right and, ever since its founding, has seen its theoretical structure remain under development. A greater understanding as regards the importance of knowledge stems from returning to works such as those by Penrose (1959) and Nelson and Winter (1982). We may correspondingly argue that while knowledge in itself represents a resource, the effective management of knowledge within a company enables extraction in ways that reach beyond all other available resources.

Furthermore, knowledge performs an important support function in supplying the bases for coordinating the mechanisms able to boost the conversion of resources into capacities. Thus, knowledge has now become perceived as the fulcral cornerstone of companies, playing an ever-increasing role in spotting and capitalizing on entrepreneurial opportunities (Andersson and Hellerstedt 2009). The rising number of analytical studies on the importance of entrepreneurship at the regional level and the characteristics of locations have together demonstrated that knowledge lies at the foundations for launching new companies (Varga 2000; Audretsch and Lehmann 2005; Riddel and Schwer 2003; Fernandes and Ferreira 2014).

C. Fernandes (✉) · J. J. Ferreira
Management and Economics Department, University of Beira Interior, Covilhã, Portugal
e-mail: jjmf@ubi.pt

P. M. Veiga
Universidade Portucalense Infante D. Henrique & NECE- Research Unit, Covilhã, Portugal
e-mail: motaveiga@curva-de-gauss.pt

M. Peris-Ortiz
Department of Business Administration, Universitat Politècnica de València, Valencia, Spain
e-mail: mperis@doe.upv.es

M. Peris-Ortiz et al. (eds.), *Knowledge, Innovation and Sustainable Development in Organizations*, Innovation, Technology, and Knowledge Management,
https://doi.org/10.1007/978-3-319-74881-8_2

According to Acs et al. (2006), entrepreneurial activities are tending to become even better in the sense of investments and are heading into new knowledge at relatively high levels simultaneous to companies, especially new firms and businesses, making recourse to the genuine source of original knowledge as generated by universities. Thus, and as had already been defined by Schumpeter (1934, 1939, 1942), entrepreneurship represents the leading driving force of leb, able to come up with the innovations that generate profits while taking on the risks inherent to such creative ventures. Innovation has correspondingly gained broad recognition as one of the main drivers of growth within a period often termed the "knowledge era" (Stough 2003; Mention 2011). Hence, in a global and increasingly competitive business environment, innovation therefore increasingly also becomes a critical factor for companies striving either to assume dominant positions (Cheng et al. 2010) or to revitalize their core competences (Hu and Hsu 2008; Kaminski et al. 2008). This furthermore perceives innovation as one of the main methods for adaptation to the ever more dynamic surrounding environment (Roberts and Amit 2003; Hua and Wemmerlov 2006; Doloreux and Melancon 2008).

However, sustainability is now presenting an increasing challenge to innovation processes, whether of the innovations themselves or of the surrounding environment due to the rising scope of environmental challenges (Sharma and Vredenburg 1998; van Kleef and Roome 2007; Huizingh 2011; Keskin et al. 2013). Hence, within this context of knowledge, innovation and sustainability, there is particular relevance in systematizing the literature hitherto published in order to identify possible future trends.

This article therefore seeks to meet this need as reflected in the literature through bibliometric analysis that focuses on the way in which these concepts (knowledge, innovation and sustainability) have interrelated over the course of time given that the respective fields of study emerge as highly dispersed and thus far lacking any systematic study of the literature in these terms.

Currently, bibliometric analysis represents the most commonly applied methodology for analysing research findings and the literature (Mutschke et al. 2011). This spans the application of quantitative and statistical analysis to academic outputs, especially articles and their respective citations (Thomsom Reuters 2008) in order to generate a vast perspective over the research activities and their impacts, particularly in terms of researchers, journals, countries and universities of origin (Hawkins 1977; Osareh 1996).

The study objective involves providing researchers into themes surrounding knowledge, innovation and sustainability with a map of the journals, authors and themes that enables a better understanding of the publications interrelated with the theme.

2.1.1 Literature Review

Schumpeter (1934, 1939, 1942) identifies the entrepreneur as the driving force behind growth and economic development. Indeed, such individuals are able to produce the innovations that ensure the return of profits while taking on the risks inherent to such "creations". According to this author, development means the introduction of new combinations of circular flows into economic life; thus, entrepreneurs bring about the introduction of innovative actions to such an extent that they may cause cyclical discontinuities in the economy. Afterwards, Drucker (1985) argues that innovation offers a powerful instrument for entrepreneurs.

According to Tidd et al. (1997), the creation of wealth by organizations comes from the processes of innovation. Innovation is thus the invention in the mode of commercialization that is to say with practical utility.

Acs and Audretsch (1988) developed an investigation adding to the existing literature new ways of examining technological change through the more direct introduction of variables that measure innovative activity, in order to determine some of its basic properties. This was done through a model that studied the degree to which innovation is affected by the characteristics of different industries and to what extent small and medium-sized enterprises respond to various stimuli. They concluded that the innovative activity of small and medium-sized enterprises responds considerably to the different economic and technological environments. Acs and Varga (2005) developed their research based on two relationships, geography and technology and entrepreneurship and technology, arguing that these relationships are key to explaining economic development. They concluded that the effects of agglomeration on technological change, or if we want innovation, are positive and statistically significant.

As argued by several authors, innovation is critical in the achievement of competitive advantage by organizations (Hu and Hsu 2008; Kaminski et al. 2008), as well as allowing companies to adhere to an increasingly complex environment (Roberts and Amit 2003; Hua and Wemmerlov 2006; Fernandes et al. 2017).

Innovation comes from the flexibility of enterprises to be able to choose different options to satisfy the desires of consumers (Banbury and Mitchell 1995) through a sustained strategy focused on the resources and capabilities of companies that allow not only to satisfy these desires today but also in the future (Wernerfelt 198; Barney 1991; Drazin and Schoonhoven 1996; Tushman and O'Reilly 1997; Souitaris 2002; Hwang 2004; Lemon and Sahota 2004).

Knowledge thus gains recognition as constituting the core foundations of companies in conjunction with a rising role in the process of identifying and leveraging entrepreneurial opportunities (Andersson and Hellerstedt 2009). Hence, we need to approach sustainability as a dynamic and layered concept attained over the course of time in fairly dichotomist manners (sustainable/non-sustainable) (Adams et al. 2016). Therefore, fundamentally, companies need to determine whether there are sustainable paths available for a specific business and, whenever not encountering them, identifying just how companies might incorporate sustainability into their activities (Nidumolu et al. 2009).

Recently, academic progress has pointed to innovation as a persuasive means of improving business sustainability (Hart 1995; Hansen et al. 2009; Dangelico and Pujari 2010; Seebode et al. 2012; Horn and Brem 2013) or even as the touchstone for all innovations given that the similarities shared by their objectives (Nidumolu et al. 2009).

2.1.2 Methodology

Data

We gathered our data from the Science Citation Index Expanded (1900–present), Social Sciences Citation Index (1956–present), Arts and Humanities Citation Index (1975–present), Conference Proceedings Citation Index – Science (1990–present) and Conference Proceedings Citation Index – Social Science and Humanities (1990–present), as compiled by the online Thomson Reuters-ISI databases that contain many thousands of academic articles and bibliographic information about the authors and their affiliations and citations. The research took place in September 2017 in the Web of Science Core Collection database involving the application of the research terms *Knowledge and Innovation* and *Sustainability* to the title, abstract or keywords and with the chronological filter set for through to the end of 2016. The search returned 716 articles with publication dates of between 1994 (1 article) and 2016 (150 articles).

As regards the statistical and analytical methods applied to the database, we carried out descriptive analysis of the articles resulting from the search and primarily making recourse to graphic methods, frequency tables and descriptive statistics (median and standard deviation).

All these procedures took place through recourse to the Microsoft Excel 2010 software (Microsoft Corporation, Washington, USA).

2.1.3 Results

Figure 2.1 presents the annual trends in the number of articles getting published on this area. The average year of publication is 2012.7 ± 3.7 and correspondingly demonstrating that this is a field of research undergoing a growth phase. Through to the year 2000, the number of articles on this topic remained only at very low levels before experiencing exponential growth after 2011 and with 2016 corresponding to the year with clearly the largest number of articles published on knowledge, innovation and sustainability (150 articles).

As regards the citations, there was an average incidence of 15.1 ± 56.9 citations in which 113 (18.3%) of the articles did not register a single citation and 201 (32.6%) received no less than 10 citations.

The three articles receiving the largest number of citations are:

1. Damschroder, L. J., Aron, D. C., Keith, R. E., Kirsh, S. R., Alexander, J. A., & Lowery, J. C. (2009). Fostering implementation of health services research findings into practice: a consolidated framework for advancing implementation science. Implementation Science, 4(50). **(1152 citations, 128 citations per year)**
2. Growth, innovation, scaling, and the pace of life in cities. Proceedings of The National Academy of Sciences of The United States of America, 104(17), 7301–7306 **(525 citations, 47.7 citations per year)**

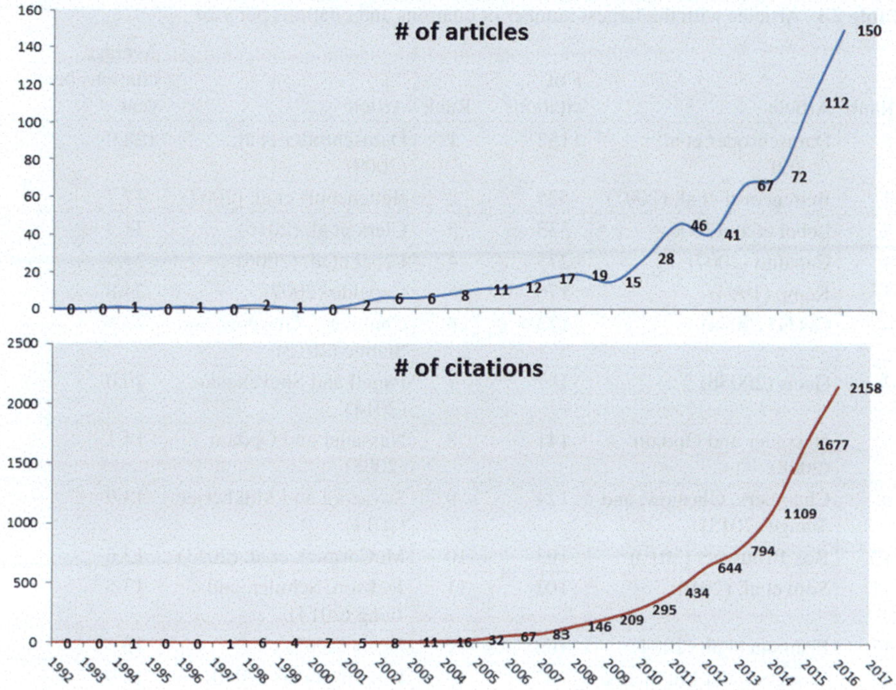

Fig. 2.1 Number of articles and citations per year

3. Lebel, L., Anderies, J. M., Campbell, B., Folke, C., Hatfield-Dodds, S., Hughes, T. P., & Wilson, J. (2006). Governance and the capacity to manage resilience in regional social-ecological systems. Ecology and Society, 11(19). **(333 citations, 34.5 citations per year)**

According to Damschroder et al. (2009), many interventions found to be effective in health service research studies fail to figure out into meaningful patient care outcomes across multiple contexts. Health service researchers recognize the need to evaluate not only summative outcomes but also formative outcomes to assess the extent to which implementation is effective in a specific setting, prolongs sustainability and promotes dissemination into other settings. Many implementation theories have been published to help promote effective implementation. However, they overlap considerably in the constructs included in individual theories, and a comparison of theories reveals that each is missing important constructs included in other theories.

Bettencourt et al. (2007) study the phenomenon how humanity has just crossed a major landmark in its history with the majority of people now living in cities. Cities have long been known to be society's predominant engine of innovation and wealth creation, yet they are also its main source of crime, pollution and disease. The inexorable trend towards urbanization worldwide presents an urgent challenge for developing a predictive, quantitative theory of urban organization and sustainable development. In this research they present empirical evidence indicating that the

Table 2.1 Articles with the largest number of citations and citations per year

Rank	Article	# of citations	Rank	Article	Average citations per year
1	Damschroder et al. (2009)	1152	1	Damschroder et al. (2009)	128.0
2	Bettencourt et al. (2007)	525	2	Bettencourt et al. (2007)	47.7
3	Lebel et al. (2006)	333	3	Clark et al. (2016)	34.5
4	Capaldo (2007)	273	4	Lebel et al. (2006)	27.8
5	Kemp (1994)	176	5	Capaldo (2007)	24.8
6	Geels (2005a)	173	6	Chambers, Glasgow, and Stange (2013)	24.8
7	Geels (2005b)	167	7	Pagell and Shevchenko (2014)	20.0
8	Nassauer and Opdam (2008)	141	8	Nassauer and Opdam (2008)	14.1
9	Chambers, Glasgow, and Stange (2013)	124	9	Sovacool and Mukherjee (2011)	13.9
10	Bos-Brouwers (2010)	103	10	McCormick et al. (2013)	13.6
11	Som et al. (2010)	102	11	Jackson, Schuler, and Jiang (2014)	13.5
12	Fishman et al. (2004)	102	12	Geels (2005a)	13.3
13	Sovacool and Mukherjee (2011)	97	13	Bos-Brouwers (2010)	12.9
14	Jenkins (2009)	95	14	Geels (2005b)	12.8
15	Smits (2002)	89	15	Som et al. (2010)	12.8

processes relating urbanization to economic development and knowledge creation are very general, being shared by all cities belonging to the same urban system and sustained across different nations and times.

The sustainability of regional development can be exploited in a useful way through several different lenses (Lebel et al. 2006). Lebel et al. (2006) argue that in situations where uncertainties and changes are fundamental characteristics of the ecological landscape and social organization, the critical factors for sustainability are resilience, the ability to cope and adapt and the conservation of sources of innovation and renewal.

Table 2.1 presents the 15 articles with the largest number of citations and the citations per year.

As regards the sources, we may report that the 496 articles returned by the search came out in 369 journals, of which 284 (77.0%) had published but a single article on this theme.

Figure 2.2 depicts the journals with the largest number of published articles, with the largest number of citations, with the highest citation average per article and with the largest number of articles easily belonging to the *Journal of Cleaner Production* (45 articles), followed by the journals *Sustainability* (16 articles) and *Ecology and Society* (10 articles). Generating the greatest number of citations was *Implementation*

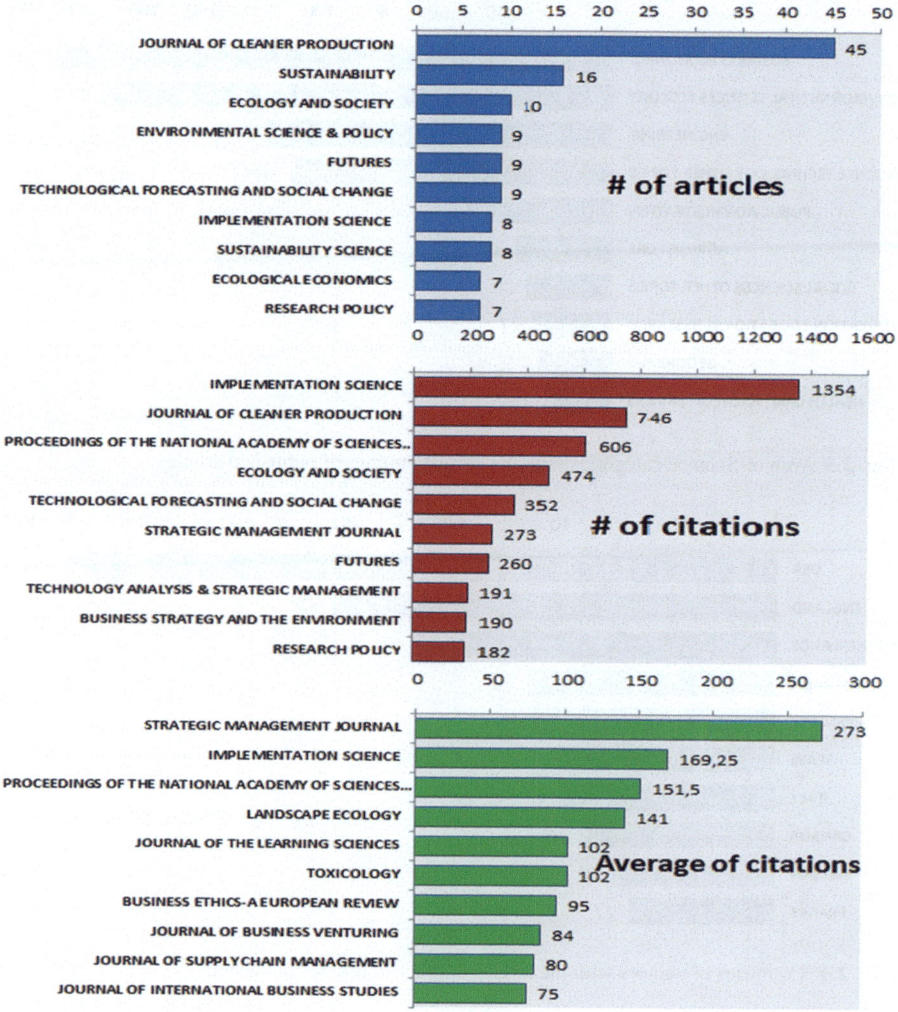

Fig. 2.2 Journals with the largest number of published articles, the largest number of citations and the highest average citations per article

Science (1354 citations), the *Journal of Cleaner Production* (746 citations) and the *Proceedings of the National Academy of Sciences of the United States of America* (606 citations). Finally, in terms of the highest average level of citations comes the *Strategic Management Journal* (273.0 citations per article) and *Implementation Science* (169.3 citations per article).

Due to the transdisciplinary nature of the field in question, we carried out analysis of the Web of Science categories (Fig. 2.3) that details the results of the categories containing the largest number of published articles. The categories business economics (190 articles; 30.8%), environmental sciences ecology (175 articles;

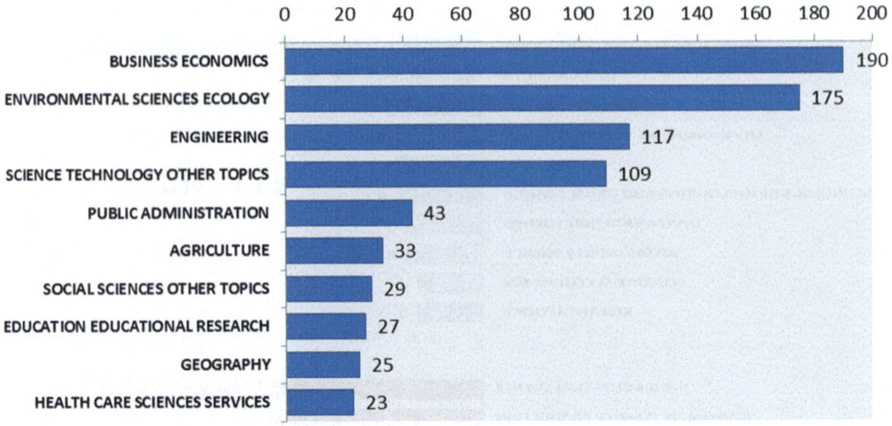

Fig. 2.3 Web of Science categories with the largest number of published articles

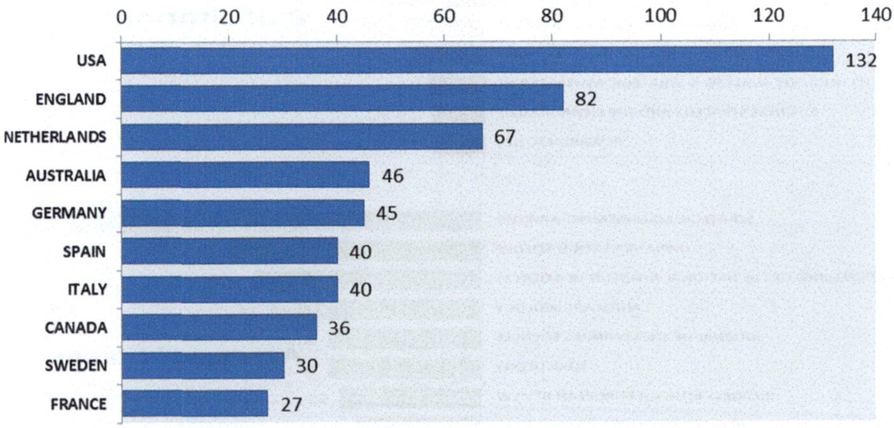

Fig. 2.4 Countries of authors with the largest number of articles published

28.4%) and engineering (117 articles; 19.0%) prevailed in terms of the number of articles published. We would reference that each article might be listed simultaneously in more than one category.

In terms of author countries of origins, of the 617 articles published in the Web of Science Core Collection database (Fig. 2.4), the USA (132 articles; 21.4%), England (82 articles; 13.3%) and the Netherlands (67 articles; 10.9%) stand out as the countries with the greatest number of articles published.

As regards author affiliation (Fig. 2.5), we may report Wageningen University and Research (16 articles) and Delft University of Technology in the Netherlands (11 articles) and the University of California System in the USA (11 articles) as the institutions with the highest levels of outputs in terms of articles published on this field.

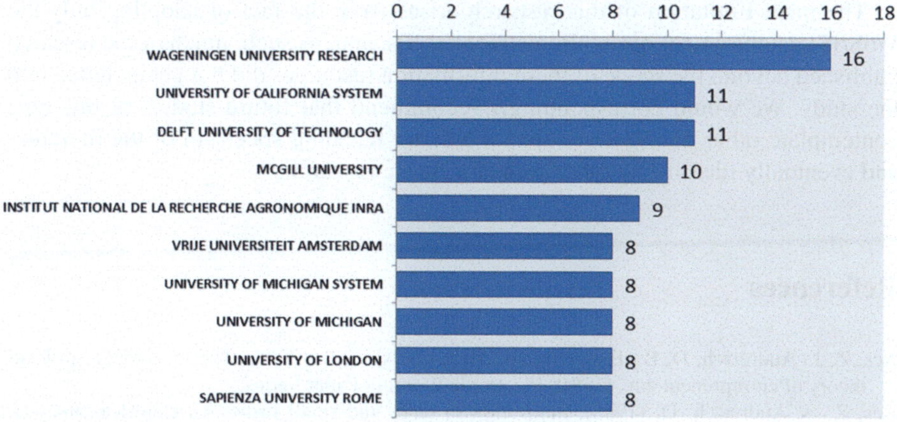

Fig. 2.5 Institutions of authors with the largest number of articles published

2.1.4 Final Considerations

Based upon the results returned, we find that the literature has emphasized how knowledge generates innovation and this has an impact on the founding of new companies and that innovation is bound up with the surrounding environment. These findings thus reveal a rising level of concern over the needs of the planet. We may furthermore report how this theme is both recent and undergoing clear expansion and hence the need for systematization and deeper analysis of the literature to grasp the scope of future trends.

We would also highlight how this theme generally appears in the literature associated with large sums of capital investments, the application of technical and scientific means to productive systems and a profound restructuring of the economy. Hence, the provision of infrastructures for such spaces stems from considerable public investment designed to make private capital more productive on the expectation that, as a consequence, the expansion in the networks and systems enables these two important results. Entrepreneurship support policies have correspondingly become the rule across diverse territorial levels even down to that of the most remote rural regions.

Over the last decade, the European Union and many OECD member states have introduced policies that deploy entrepreneurship as a core tool for rural development. Similarly, there is both growing demand and interest in setting up and running new business with this proving a key factor within the process of knowledge, innovation and the sustainable development of economies. What are the tools, models, scales and mechanisms that companies should adopt and explore in order to be able to continue to become increasingly competitive and innovative and follow their simultaneously sustainable growth policies? These are some of the main challenges faced by researchers, entrepreneurs and policy makers. Future lines of research should also try and open up and explore paths in this area and thereby contribute towards greater knowledge about this theme.

The main limitation of this research arises from the fact of adopting only the WoS/ISI databases for our bibliometric analysis, and, as such, much of the research published beyond the scope of these information resources did not get included into the study. We would correspondingly recommend that future studies of this type contemplate other databases to span a broader reaching spectrum of the literature and eventually identifying other paths for future research.

References

Acs, Z. J., Audretsch, D. B., Braunerhjelm, P., & Carlsson, B. (2006). The knowledge spillover theory of entrepreneurship, CESIS Electronic Working Paper Series.

Acs, Z., & Audretsch, D. (1988). Innovation in large and small firms: An empirical analysis. *American Economic Review, 78*(4), 678–690.

Acs, Z., & Varga, A. (2005). Entrepreneurship, agglomeration and technological change. *Small Business Economics, 31*, 630–640.

Adams, R., Jeanrenaud, S., Bessant, J., Denyer, D., & Overy, P. (2016). Sustainability oriented innovation: A systematic review. *International Journal of Management Reviews, 18*, 180e205.

Andersson, M., & Hellerstedt, K. (2009). Location attributes and start-ups in knowledge-intensive business services. *Industry & Innovation, 16*(1), 103–121.

Audretsch, D. B., & Lehmann, E. E. (2005). Does the knowledge spillover theory of entrepreneurship hold for regions? *Research Policy, 34*(8), 1191–1202.

Banbury, C. M. & Mitchell, W. (1995). The effect of introducing important incremental innovations on market share and business survival. *Strategic Management Journal (summer special issue), 16*, 161–182.

Barney, J. (1991). Firm resources and sustained competitive advantage. *Journal of Management, 17*(1), 99–120.

Bettencourt, L. M. A., Lobo, J., Helbing, D., Kuehnert, C., & West, G. B. (2007). Growth, innovation, scaling, and the pace of life in cities. *Proceedings of the National Academy of Sciences of the United States of America, 104*(17), 7301–7306.

Bos-Brouwers, H. E. J. (2010). Corporate sustainability and innovation in SMEs: Evidence of themes and activities in practice. *Business Strategy and the Environment, 19*(7), 417–435.

Capaldo, A. (2007). Network structure and innovation: The leveraging of a dual network as a distinctive relational capability. *Strategic Management Journal, 28*(6), 585–608.

Chambers, D. A., Glasgow, R. E., & Stange, K. C. (2013). The dynamic sustainability framework: Addressing the paradox of sustainment amid ongoing change. *Implementation Science, 8*, 117.

Cheng, C., Lai, M., & Wu, W. (2010). Exploring the impact of innovation strategy on R&D employees job satisfaction: A mathematical model and empirical research. *Technovation, 30*, 459–470.

Clark, W. C., Tomich, T. P., van Noordwijk, M., Guston, D., Catacutan, D., Dickson, N. M., & McNie, E. (2016). Boundary work for sustainable development: Natural resource management at the Consultative Group on International Agricultural Research (CGIAR). *Proceedings of the National Academy of Sciences of the United States of America, 113*(17), 4615–4622.

Damschroder, L. J., Aron, D. C., Keith, R. E., Kirsh, S. R., Alexander, J. A., & Lowery, J. C. (2009). Fostering implementation of health services research findings into practice: A consolidated framework for advancing implementation science. *Implementation Science, 4*, 50.

Dangelico, R. M., & Pujari, D. (2010). Mainstreaming green product innovation: Why and how companies integrate environmental sustainability. *Journal of Business Ethics, 95*, 471–486.

Doloreux, D., & Melancon, Y. (2008). On the dynamics of innovation in Quebec's coastal maritime industry. *Technovation, 28*, 231–243.

Drazin, R. & Schoonhoven, C. B. (1996). Community, population, and organization effects on innovation: A multilevel perspective. *Academy of Management Journal, 39*(5), 1065–1083.

Drucker, P. (1985). *Innovation and Entrepreneurship – Practice and Principals*. New York: Harper & Row.

Fernandes, C., & Ferreira, J. (2014). Knowledge spillovers: Cooperation between universities and KIBS. *R&D in Management, 43*(5), 461–472.

Fernandes, C., Veiga, P., Peris-Ortiz, M., & Rueda-Armengot, C. (2017). What impact does innovation and sustainable entrepreneurship have on competitiveness? *Journal of Social Ecology and Sustainable Development, 8*(3), 56–66.

Fishman, B., Marx, R. W., Blumenfeld, P., Krajcik, J., & Soloway, E. (2004). Creating a framework for research on systemic technology innovations. *Journal of the Learning Sciences, 13*(1), 43–76.

Geels, F. W. (2005a). Processes and patterns in transitions and system innovations: Refining the co-evolutionary multi-level perspective. *Technological Forecasting and Social Change, 72*(6), 681–696.

Geels, F. W. (2005b). The dynamics of transitions in socio-technical systems: A multi-level analysis of the transition pathway from horse-drawn carriages to automobiles (1860–1930). *Technology Analysis & Strategic Management, 17*(4), 445–476.

Hansen, E. G., Grosse-Dunker, F., & Reichwald, R. (2009). Sustainability innovation cube: A framework to evaluate sustainability-oriented innovations. *International. Journal of Innovation and Management, 13*, 683e713.

Hart, S. L. (1995). A natural-resource-based view of the firm. *Academy of Management Review, 20*, 986e1014.

Hawkins, D. T. (1977). Unconventional uses of on-line information retrieval systems: On-line bibliometric studies. *Journal of the American Society for Information Science, 28*(1), 13–18.

Horn, C., & Brem, A. (2013). Strategic directions on innovation management: A conceptual framework. *Management Research Review, 36*, 939e954.

Hu, J., & Hsu, Y. (2008). The more interactive, the more innovative? A case study of South Korean cellular phone manufacturers. *Technovation, 28*, 75–87.

Hua, S., & Wemmerlov, U. (2006). Product change intensity, product advantage, and market performance: An empirical investigation of the PC industry. *Journal of Product Innovation Management, 23*, 316–329.

Huizingh, E. K. R. E. (2011). Open innovation: State of the art and future perspectives. *TECHNOVATION, 31*(1, SI), 2–9. http://doi.org/10.1016/j.technovation.2010.10.002.

Hwang, A. (2004). Integrating technology marketing and management innovation. *Research Technology Management, 47*(4), 27–31.

Jackson, S. E., Schuler, R. S., & Jiang, K. (2014). An aspirational framework for strategic human resource management. *Academy of Management Annals, 8*(1), 1–56.

Jenkins, H. (2009). A `business opportunity' model of corporate social responsibility for small- and medium-sized enterprises. *Business Ethics-A European Review, 18*(1), 21–36.

Kaminski, P., de Oliveira, A., & Lopes, T. M. (2008). Knowledge transfer in product development processes: A case study in small and medium enterprises (SMEs) of the metal mechanic sector from São Paulo, Brazil. *Technovation, 28*, 29–36.

Kemp, R. (1994). Technology and the transition to environmental sustainability – the problem of technological regime shifts. *Futures, 26*(10), 1023–1046.

Keskin, D., Diehl, J. C., & Molenaar, N. (2013). Innovation process of new ventures driven by sustainability. *Journal of Cleaner Production, 45*, 50–60.

Lebel, L., Anderies, J. M., Campbell, B., Folke, C., Hatfield-Dodds, S., Hughes, T. P., & Wilson, J. (2006). Governance and the capacity to manage resilience in regional social-ecological systems. *Ecology and Society, 11*(1), 19.

Lemon, M. & Sahota, P. (2004). Organizational culture as a knowledge repository for increased innovative capacity. *Technovation, 24*, 483–498.

McCormick, K., Anderberg, S., Coenen, L., & Neij, L. (2013). Advancing sustainable urban transformation. *Journal of Cleaner Production, 50*, 1–11.
Mention, A. (2011). Co-operation and co-opetition as open innovation practices in the service sector: Which influence on innovation novelty? *Tehnovation, 31*, 44–53.
Mutschke, P., Mayr, P., Schaer, P., & Sure, Y. (2011). Science models as value-added services for scholarly information systems. *Scientometrics, 89*(1), 349–364.
Nassauer, J. I., & Opdam, P. (2008). Design in science: Extending the landscape ecology paradigm. *Landscape Ecology, 23*(6), 633–644.
Nelson, R. R., & Winter, S. G. (1982). *An evolutionary theory of economic change*. Cambridge, MA: Harvard University Press.
Nidumolu, R., Prahalad, C. K., & Rangaswami, M. R. (2009). Why sustainability is now the key driver of innovation. *Harvard Business Review, 87*, 56e64.
Osareh, F. (1996). Bibliometrics, citation analysis and co-citation analysis: A review of Literature I. *Libri, 46*(3), 149–158.
Pagell, M., & Shevchenko, A. (2014). Why research in sustainable supply chain management should have no future. *Journal of Supply Chain Management, 50*(1), 44–55.
Penrose, E. (1959). *The theory of the growth of the firm*. Oxford: Oxford University Press.
Riddel, M., & Schwer, R. (2003). Regional innovative capacity with endogenous employment: Empirical evidence from the US. *The Review of Regional Studies, 33*(1), 73–84.
Roberts, P., & Amit, R. (2003). The dynamics of innovative activity and competitive advantage: The case of Australian retail banking, 1981 to 1995. *Organization Science, 14*(2), 107–122.
Schumpeter, J. A. (1934). *The theory of economic development*. Cambridge, MA: Harvard University Press.
Schumpeter, J. A. (1939). *Business cycles*. New York: McGraw – Hill.
Schumpeter, J. A. (1942). *Capitalism, socialism and democracy*. New York: Harper & Row.
Seebode, D., Jeanrenaud, S., & Bessant, J. (2012). Managing innovation for sustainability. *R&D Management, 42*, 195e206.
Sharma, S., & Vredenburg, H. (1998). Proactive corporate environmental strategy and the development of competitively valuable organizational capabilities. *Strategic Management Journal, 19*, 729e753.
Smits, R. (2002). Innovation studies in the 21st century: Questions from a user's perspective. *Technological Forecasting and Social Change, 69*(9), 861–883.
Som, C., Berges, M., Chaudhry, Q., Dusinska, M., Fernandes, T. F., Olsen, S. I., & Nowack, B. (2010). The importance of life cycle concepts for the development of safe nanoproducts. *Toxicology, 269*(2–3, SI), 160–169.
Souitaris, V. (2002). Technological trajectories as moderators of firm-level determinants of innovation. *Research Policy, 31*, 877–898.
Sovacool, B. K., & Mukherjee, I. (2011). Conceptualizing and measuring energy security: A synthesized approach. *Energy, 36*(8), 5343–5355.
Stough, R. (2003). Strategic management of places and policy. *Annals of Regional Science, 37*(1), 179–201.
Thomsom Reuters (2008). Using bibliometrics: A guide to evaluating research performance with citation data (Retrieved). Retrieved from http://ip-science.thomsonreuters.com/m/pdfs/325133_thomson.pdf
Tidd, J., Bessant, J., & Pavitt, K. (1997). *Managing Innovation*. Chichester: Wiley.
Tushman, M., & O'Reilly, C. (1997). Winning through innovation: A practical guide to leading organizational change and renewal. III. Boston: Harvard Business School Press.
van Kleef, J. A. G., & Roome, N. J. (2007). Developing capabilities and competence for sustainable business management as innovation: A research agenda. *Journal of Clean Production, 15*, 38e51.
Varga, A. (2000). Local academic knowledge spillovers and the concentration of economic activity. *Journal of Regional Science, 40*, 89–309.

Chapter 3
Project Management Office in the Nongovernmental Organization as a Driver of Sustainable Competitive Advantage: A Dynamic Capabilities Approach

Adonai J. Lacruz, Everton A. Cunha, Ralf L. de Moura, and Marcos P. V. de Oliveira

3.1 Introduction

Nongovernmental organizations (NGOs) generally develop their actions through projects in the fulfillment of their institutional missions (Lacruz and Cunha ahead of print; Diallo and Thuillier 2004). Some have found an alternative to improve project performance in the implementation of organizational units called the project management offices (PMOs) that achieve and sustain competitive advantage (Golini et al. 2014).

The historical trajectory of NGO formation—particularly in the context of Latin America—indicates the motivation for NGOs to adopt managerial practices inspired by for-profit organizations (Alvarez 2009; Silva 2010; Tenório 1999). Although its origin is attributed to a philanthropic model, the dynamism of its business environment, and consequent increase in competitive rivalry, encourages

A. J. Lacruz (✉)
Federal Institute of Education, Science and Technology of Espírito Santo (Ifes), Viana, Brazil

Post-Graduation Program in Administration, Economics and Legal Sciences Center, Federal University of Espírito Santo (UFES), Vitória, Brazil
e-mail: adonail.lacruz@ifes.edu.br

E. A. Cunha
Business Administration, FUCAPE Business School, Vitória, Brazil

R. L. de Moura · M. P. V. de Oliveira
Post-Graduation Program in Administration, Economics and Legal Sciences Center, Federal University of Espírito Santo (UFES), Vitória, Brazil
e-mail: marcos.p.oliveira@ufes.br

M. Peris-Ortiz et al. (eds.), *Knowledge, Innovation and Sustainable Development in Organizations*, Innovation, Technology, and Knowledge Management,
https://doi.org/10.1007/978-3-319-74881-8_3

NGOs to migrate to a corporate model. For Alvarez (2009), this is called the "NGO-ization" of social movements.

Similar to the stock market, there is a donation market (Glaeser 2002) in which resources for the development of NGO projects are offered by public and private organizations. These are solicited by NGOs, usually by submitting project proposals in response to public policies' calls. As a consequence, more intense competition for resources exerts force for the professionalization of NGO management processes (Alvarez 2009; Silva 2010; Tenório 1999), such as the adoption of project management practices (Lacruz and Cunha ahead of print) in the search for greater efficiency, effectiveness, and transparency in their actions.

The dynamic capabilities approach points to the importance of incorporating the dynamism of the environment into the determination of competitive advantage. The way organizations react to the dynamism of the environment, whether through routines and processes (Teece 2007) or other capabilities (McKelvie and Davidsson 2009), allows organizations to reconfigure their capabilities, leading to new configurations and sustainable competitive advantage.

There is much literature demonstrating the positive influence of project management practices on project performance (e.g., Joslin and Müller 2015; Liu 2015) and, in the last decade, the analysis of the possible PMO moderating role in this context (e.g., Jalal and Koosha 2015; Dai and Wells 2004). However, a research gap remains, indicating the need for studies that adopt NGOs as an object of research and, especially, the analysis from the perspective of dynamic capability (Teece et al. 1997). These studies should have an analytical lens that may reveal new meanings to the role of the PMO in NGOs.

In addition, these studies approach the discussion using a transversal cut as a research strategy, which limits the disclosure of the possible role of PMO moderator, and markedly from observations in private for-profit organizations (e.g., Yazici 2009).

In response to the identification of this research gap in the literature, this study investigates the PMO's role in NGO internal project performance (triple restriction) under the lens of dynamic capabilities. It adopts a longitudinal cut (3 years) through an ex post facto research (Chapin 1947) using the difference-in-differences (DID) technique in a set of six projects of the same NGO operating in Brazil. Thus, the sample will be composed of four proportional groups, control group (1) before and (2) after the implantation of the PMO and treatment group (3) before and (4) after the PMO implantation, so that the evaluation of the impact of the possible role of moderating the PMO will be done by analyzing the double difference. Thus, the question of this research objectively presents itself: *is the performance of the projects in the period after the implementation of the PMO higher than in the period prior to implementation?*

3.2 Background

The perspective of dynamic capacities allows us to explain how organizations are renewed in response to environmental changes (Teece et al. 1997). Its conceptual structure, as emphasized by Augier and Teece (2008), is based mainly on the combination of conceptual elements of resource-based view (Barney 1986; Penrose 1959), transaction cost economics (Williamson 1975, 1985), and the firm's neo-Schumpeterian view (Nelson and Winter 1982), which combine with the ideas of the innovation dynamics proposed by Schumpeter (1934, 1942).

The seminal concept of dynamic capabilities, in the initial definition proposed by Teece et al. (1997), refers to the organization's ability to integrate, construct, and reconfigure external and internal competencies in regard to the dynamism of the environment (Teece et al. 1997).

The concept of dynamic capabilities evolves on a varied basis. Although there is a relationship between these definitions, it is not uncommon for each author to emphasize one or more particular aspects of dynamic capabilities. In general terms, the definitions focus on the set of behaviors, abilities, and capabilities (e.g., McKelvie and Davidsson 2009), processes (e.g., Teece 2007), and organizational learning mechanisms (e.g., Crossan et al. 1999).

At the core of the concept of dynamic capabilities is the organization's ability to reconfigure its resources. This process will impact organizational performance. According to Winter (2003) and Zollo and Winter (2002), for a capability to be considered dynamic, the organization must be able to use it repeatedly and reliably. Thus, ad hoc solutions are not considered dynamic capabilities.

Also, for Teece (2007), dynamic capacity is the ability of the organization to identify environmental opportunities and threats, as well as respond efficiently to changes in the environment, from the adaptation/renewal and exploration of its internal and external competences, gaining lasting competitive advantage "... enterprise's capacity to successfully innovate and capture sufficient value to deliver superior long-term financial performance" (p. 1320). In this sense, not all the organizational responses given to environmental mutations, propelling threats and opportunities, can be considered manifestations of dynamic capabilities (Teece 2007), since some of them are not capable of improving the competitive performance of the organization.

Most NGO initiatives are conducted in the form of projects (Lacruz and Cunha ahead of print; Diallo and Thuillier 2004), which underscores their importance to these entities. As an effort to make managing their projects more effective (Pellegrinelli and Garagna 2009; Jalal and Koosha 2015), organizations have used the strategy of establishing and integrating a unit that is considered innovative and highly effective into their organizational structures (Alexandrova et al. 2015), typically known as PMO, present in the context of NGOs (Golini et al. 2014).

This study investigates the possible moderation exerted by PMO in the specific relationship between dynamic capabilities and project management performance. There is much literature demonstrating that project management has a positive influence on project performance (e.g., Joslin and Müller 2015; Liu 2015) and the PMO's moderating role in this context in the last 10 years (e.g., Jalal and Koosha 2015; Dai and Wells 2004). However, few reports have adopted NGOs as an object of research (e.g., Golini et al. 2014) and especially the analysis from the perspective of dynamic capabilities that has an analytical lens that can reveal new meanings to the eventual PMO's NGOs and for which empirical studies are rare (e.g., Biesenthal et al. 2012).

The motivation for NGO adoption of project management practices in general, and the implementation of PMOs in particular, stems in part from the dynamism of the historical trajectory of NGO formation, particularly in the context of Latin America (Alvarez 2009; Silva 2010; Tenório 1999). This has encouraged NGOs to migrate from a fundamentally philanthropic model to the corporate, titled by Alvarez (2009) as "NGO-ization" of social movements.

As a consequence, the corporate model brings together partnerships that trigger temporary organizational arrangements, in line with Turner and Muller's (2003) vision—of which projects are temporary organizations—to coordinate activities resulting from partnerships with business, government, foundations, and business associations and other NGOs (arising from the process of raising funds in the donation market). Thus, the more intense competition for public and private resources pushes NGOs to professionalize their processes and management models (Alvarez 2009; Silva 2010), adopting techniques and methods of project management (Lacruz and Cunha ahead of print) with the purpose of becoming competitive.

The PMO is usually seen as a center of excellence that implements practices and standardizes project-related governance processes, facilitating the sharing of resources, methodologies, tools, and techniques among the various projects (Project Management Institute [PMI] 2013). Its main objective is to improve project management effectiveness (Stanleigh 2006). The positive effects of PMO on project results and project portfolio management have already been demonstrated by several studies (e.g., Dai and Wells 2004; Desouza and Evaristo 2006; Unger et al. 2012).

There is no agreement on the typology of PMOs. Empirical research such as that of Hobbs and Aubry (2008) have demonstrated weaknesses in the definition of responsibilities and that PMO structures are the main reasons for the difficulty of standardizing and typing PMOs. According to the PMI (2013), PMOs should be classified according to their form of performance that varies in relation to the degree of control and influence they have in the organization. The PMO's functions can be support, functioning as a consultancy through the creation of templates, best practices, training, and lessons learned for other projects, and control, requiring compliance through the adoption of methodologies and the verification of compliance by the projects and the board, where the PMO takes control of the projects through the direct management of the projects.

Bates (1998) argues that the PMO has the support and leadership function. Block and Frame (1998) affirm that the PMO has the function of supporting projects by reducing administrative functions, consulting, guidance, developing standards and methods, and training and support to project teams. Kwak and Dai (2000) understand that the PMO's role depends on the size of the organization and adds the development of historical project data and human resource assistance to PMO functions.

The variety of responsibilities and functions of PMOs shows that the scope of their work is unclear and their characteristics depend on factors such as organizational characteristics (Hobbs and Aubry 2008), specific organizational requirements (Bates 1998), organizational objectives (Kwak and Dai 2000), and the portfolio of projects in the organization (Cooper et al. 1997). Other researches (Aubry et al. 2010; Hurt and Thomas 2009) analyze the frequent transformation of the PMO and its patterns of change, due to the change in said factors. Cooper et al. (1997) argue that as the list of active projects in the organization (portfolio) is regularly updated, it makes PMO's portfolio management a dynamic process, thus requiring dynamic capabilities.

Dynamic capabilities[1] are made up of three main capacities: (1) to sense and shape opportunities and threats, (2) to seize opportunities, and (3) to maintain competitiveness by improving, combining, protecting, and, if necessary, transforming and reconfiguring of the organization's intangible and tangible assets (Teece 2009).

In this context, PMOs are transforming organizational units where their roles and functions evolve on a continuum in relation to environmental change. Petit and Hobbs (2010) understand that the organization's project governance is shrouded in routine as governance is systematic over time. Regular changes in the project portfolio, for example, generate new needs and opportunities (sensing) that make organizations, through PMOs, develop dynamic capabilities by adapting to changing needs and opportunities (transforming and reconfiguring) (Teece 2009).

Adapting to changes in the project portfolio involves a combination of organizational structures, processes, and people involved in the seizing process (Killen et al. 2008). Project portfolio management involves a number of decision-making bodies and rules that are usually defined in the governance structure of the entity. In practice, the entity may need to reconfigure and reallocate existing resources and potentially develop new resources, which involves changing the enterprise's routines (Petit and Hobbs 2010).

According to Petit and Hobbs (2010), project, portfolio, and program governance, which is a function of PMOs, has a dynamic ability to perceive opportunities and threats, seize opportunities, and maintain competitive conditions by enhancing, combining, protecting, and even reconfiguring business assets.

Killen et al. (2008) further argue that establishing a holistic view on decision protocol selection is also part of the routine of PMOs, aligning with organizational strategies and delivering the best results, as well as learning activities and organizational skills that ensure the dynamism and responsiveness to the changing environment, which denotes its dynamic capacity. In this context, according to Petit and Hobbs (2010), the selection of decision-making protocols refers to decision-making by project managers and committees, who are created to manage and decide on the different components of the portfolio.

Thus, the PMO can be considered an organizational unit with transformative aptitude focused on projects, capable of providing organizations that use projects as a way of operating their strategies, particularly NGOs, with dynamic capabilities. The following research hypothesis follows: *H1, the PMO positively impacts the internal performance of the projects.*

[1]According to Teece (2009, pp. 87–88) "The particular (nonimitability) capacity business enterprises possess to shape, reshape, configure, and reconfigure assets only to respond to changing technologies and markets and escape the zeroprofit condition. Dynamic capabilities relate to the enterprise's ability to sense, seize, and adapt in order to generate and exploit internal and external enterprise-specific competences, and to address the enterprise's changing environment."

3.3 Method and Empirical Context

This study shows the potential impact of PMO on the performance of NGO projects, through an ex post facto study (Chapin 1947), using the DID technique through multiple linear regression (Ashenfelter and Card 1985) in a set of six projects of the Brazilian environmental NGO Instituto Terra (http://www.institutoterra.org/).

In the context of this study, ex post facto researches are appropriate in the analysis of natural experiments (Chapin 1947) and the evaluation of the possible impact of the implementation of PMO (exogenous event) on the performance of projects (variable explained). The DID technique, in turn, is appropriate in the analysis of control and treatment of groups in natural experiments, since it aims to isolate the impact of the exogenous event by the double difference (Ashenfelter and Card 1985). This means to extract the "pure" effect of the natural experiment on the dependent variable, ceteris paribus. Mathematically the DID method can be represented with the following equation:

$$\delta = \left(y_{T1} - y_{C1}\right) - \left(y_{T0} - y_{C0}\right)$$

Being:

δ = estimate from the difference in differences
y = mean of outcome of interest for each period and group
T and C = treatment group and control group
1 and 0 = period after and before treatment

The DID technique uses two subtractions: the first refers to the difference between the means of the variable outcome between the treatment and control groups for the periods before and after treatment, and the second refers to the difference of the initially calculated differences. This is the origin of the term difference in differences. Thus, δ is the estimate of the impact of the natural experiment on the variable to be explained. In order to make the DID technique clearer, an analytical scheme is shown in Fig. 3.1.

The main assumption of this technique is that the temporary course of the variable outcome for the control group represents what would occur with the treatment group if it had not been exposed to the exogenous event. This assumption usually cannot be verified directly in the data, but an indication of its validity appears when the trajectories of the two groups are similar in the period previous to the observed event. The essential concept is that if the trajectories resemble each other during the period before treatment, then it seems acceptable to think that the evolution of the control group after the event represents what would happen to the treatment group in the nontreatment situation in a permissible way.

In Fig. 3.2 the research project is illustrated from the classic notation system of Campbell and Stanley (1979).

The data for our study were collected in quarterly performance reports of 6 projects over 3 years, thus totaling 36 observations. The projects were classified into

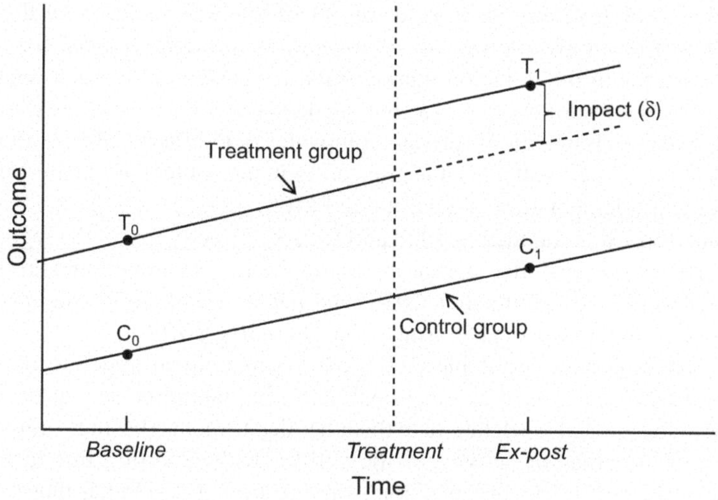

Fig. 3.1 Difference in differences

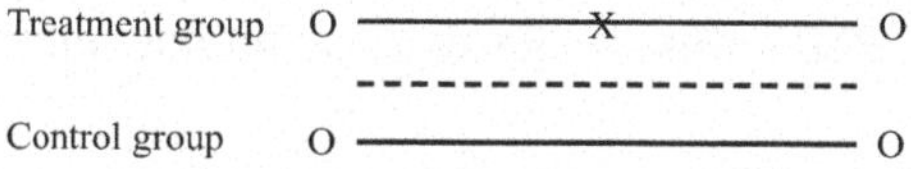

Fig. 3.2 Research project. (*Note.* X: exposure to an experimental event. O: measurement record. ____: temporal order. -----: groups not equivalent by random designation)

four proportional groups: control group before and after the implementation of the PMO and the treatment group before and after the implementation of the PMO. This was done in order to evaluate the possible impact of the PMO by the analysis of the double difference.

It is added that the Project Performance Indicator (PPI) was used as a proxy for project performance quality, composed of the simple arithmetic mean of the performance indicators in cost, time, and scope. The PPI is an index number that varies from 0 to 1, and the higher the PPI, the better the project performance. The DID method was implemented using multiple linear regression, processed in software R version 3.3.3 (R Development Core Team 2017). Its mathematical representation can be expressed as follows:

$$y = \beta_0 + \beta_1 dG + \beta_2 dT + \delta(dG.dT) + \mu$$

Being:

dG = dummy variable for the groups (control and treatment)
dT = dummy variable for the time (before and after treatment)
$dG.dT$ = interaction term
δ = coefficient of interest

The choice of Instituto Terra as an object of study is justified by three main reasons: (1) it develops a large part of its actions through projects, for which it obtains resources in the donation market (Glaeser 2002); (2) it has a relationship with a diverse set of donors from various segments of society, such as government agencies, business foundations, private companies, and other NGOs, both national and international; and finally, (3) it is pioneering in the context of the third Brazilian sector, in the implementation of a corporate PMO.

Instituto Terra was founded in 1998 and focuses its activities in the region of the Brazilian Atlantic Forest. It is currently among the largest environmental NGOs in Brazil (Análise Gestão Ambiental, 2015) and has received worldwide recognition for its work on the protection of springs (Gazeta online 2016).

Before the implantation of the PMO, the organizational structure of Instituto Terra could be characterized as functional. After implantation, as matrix, its PMO nucleated at the tactical level, lateral to the other departments (business units divided according to the areas of activity of the entity), with reporting line to the main executive of the entity. The organization's organizational design has thus combined an integrated functional structure with departmental criteria for temporary project units (covenant period).

The implementation of the PMO began in 2009 after reviewing the entity's strategic planning. The PMO was designed as a guiding unit with the following functions:

- Development, implementation, and support of methodology and project management software
- Preparation of project proposals to be submitted to potential donors
- Control of projects and definition and monitoring of performance indicators
- Development of operational process assets (e.g., workflow)
- Sharing of performance reports and documents
- Program and portfolio management

Especially since 2014 Instituto Terra has been systematically expanding the scope of its main project, called Olhos d'Água (Leitão 2016). This expansion has led Instituto Terra to have remote teams in several municipalities in the states of Minas Gerais and Espírito Santo, where the project is concentrated. The role of the PMO in this process of expansion of this project (still in progress) was absolutely important, as reported by the former Executive Director of Instituto Terra, who was responsible for implementing the project management methodology (Lacruz 2014) and the PMO (Lacruz 2015) in the entity. According to him it was necessary, considering the relationship with various donors of different sizes, legal natures, and nationalities, to establish partnerships with different stakeholders. Working with remote teams and, due to the geographic scope of the project, developing common operational protocols and making a virtual and integrated management of all areas of the project were also necessary. Additionally, establishing a road map with guidelines for defining the speed and geographical limits of each stage of the process of project expansion is also important.

Table 3.1 Global performance indicator by group and period

	Control group		Treatment group	
Descriptive statistics	Before treatment	After treatment	Before treatment	After treatment
Mean	0.81	0.77	0.79	0.93
Minimum	0.76	0.72	0.74	0.87
Maximum	0.83	0.81	0.84	0.97
Standard deviation	0.04	0.05	0.05	0.05

All this reveals the relevance of the PMO of the Instituto Terra as an object of study. The following is an analysis of the possible impact of the PMO on project performance.

3.4 Findings and Discussion

Before starting the DID procedures, the studied variable was characterized in order to broaden the understanding of the results (Table 3.1).

The examination of Table 3.1 reveals that the projects that make up the control group have, on average, lower overall performance than those that make up the treatment group after PMO implantation and greater dispersion of data (coefficient of variation of 6.5% and 5.4%, respectively). It also shows that before the PMO implantation, their averages were higher, and there was less dispersion (coefficient of variation of 4.9% and 6.3%, respectively).

As it was pointed out, the main assumption of the DID technique is that the trajectories of the two groups are similar in the period before the treatment. It can then be assumed that the temporary course of the outcome variable for the control group represents what would occur with the treatment group if they had not been exposed to treatment. In this thread, it was verified by the one-way ANOVA test that the hypothesis of equality between the treatment and control groups before the PMO implantation could not be rejected (p-value = 0.382). This finding provided support for the DID technique (see "Appendix").

The model used to verify the possible impact of the PMO on the performance of the projects through the DID technique was the multiple linear regression. The variable was the PPI, which is an index number that varies from 0 to 1. The closer the variable is to 1, the better the performance of the project will be.

In order to process the regression, dummy variables were created for the period before (0) and after (1) the implementation of the PMO (time_dummy) and for the control (0) and treatment groups (1) (groups_dummy). The other variable created was the interaction term (Mod) by the product of the two dummies.

The coefficient for the Mod variable, which is an estimate of the difference in differences, was statistically significant (p-value = 4.96e−07) and had positive impact (β = 0.173) of the PMO implementation in the project performance (see "Appendix"). So, there is evidence to refute the null hypothesis that the internal performance of the projects after the PMO implementation has not been higher than the period before its implementation.

Thus, the positive impact of the PMO on the performance of projects of the NGO under study is shown in the period considered. This study argues that PMO can be considered as an organizational unit, focused on projects, with characteristics that allow it to operate as a driver in obtaining a sustainable competitive advantage, since it has the capacity to transform project management capabilities into dynamic capabilities, to identify and/or create opportunities, as well as to mitigate and/or eliminate critical threats, providing superior performance to the organization.

Capabilities can become dynamic for two main reasons, particularly the PMO of the NGO under study (remembering that different PMOs may have different roles and functions) when compared to the characteristics of the dynamic capacities. First, the development and deployment of routines arise for the selection of decision-making protocols. This understanding converges with the understanding of Petit and Hobbs (2010), for whom the selection of decision protocols (which are part of PMO functions), aligning with organizational strategies as well as organizational learning activities, contributes to the dynamism and responsiveness to the changing environment. It is also aligned with Biesenthal et al. (2012), when they propose that dynamic capacities are used at various project levels to reconfigure the existing project management capabilities. At the same time, the findings of this project complement these studies, as it comes from the PMO impact assessment on objective project performance metrics.

It is important to point out that the PMO, the object of this analysis, was devoted to a set of non-exhaustive organizational functions that enabled better elaborated decision-making mechanisms through a centralized and coordinated vision offering an environment of greater adaptability and dynamism in the individual actions of each project and in the inter-project relations. It is pertinent to expect that the addition of new functions to the PMO can further amplify its effects on project performance and, consequently, on the competitiveness of the studied NGOs.

It is suggested that, by focusing project management on a dedicated organizational unit, it is possible to establish a holistic view on the selection of decision protocols (Killen et al. 2008) to respond to environmental changes more quickly and flexibly, leading the projects to better levels of performance.

The proposed analytical framework suggests that PMO contributions to internal project performance may be associated with processes that are used in multiple projects. PMOs tend to provide organizations with past experience, allowing organizations to transfer knowledge from one project to another, replicating successful actions and seeking alternatives to unsuccessful actions, by developing historical data from past projects and, in accordance with the proposals of Kwak and Dai (2000) and Julian (2008), gaining a sustainable competitive advantage through organizational learning (Crossan et al. 1999).

Thus, it is argued that the PMO contributed to the NGO being able to perceive opportunities and threats (sense), avail opportunities (seize), and maintain competitive conditions for the continuous improvement of business assets (transforming and reconfiguring). Therefore, it does not constitute itself in a dynamic capacity, recognizing that it is assumed as an internal strategic resource of the entity (Lacruz and Cunha ahead of print) or that its functions are dynamic capacities. This contrasts with the argument of Petit and Hobbs (2010) and Biesenthal et al. (2012) that says their functions empower project management capabilities to become dynamic capabilities.

In this sense, the PMO can be understood as a resource capable of providing the organization with dynamic capabilities, a source of sustainable competitive advantage for the studied NGO.

3.5 Conclusions

A project considered as a temporary effort to create a unique product or service has unique characteristics, challenges, and requirements. Unlike managing production processes, as each project is unique, organizations need to adapt their capabilities according to the nature and specificity of each project. Additionally, for many organizations, including NGOs, managing multiple projects simultaneously requires a considerable effort that often fails to achieve performance results in cost, time, or scope.

Although a wide range of studies uses the theoretical underpinnings of dynamic capabilities, there is a lack of studies involving their relationship with project management, especially in the context of NGOs. Consequently, this study contributes to a better understanding of dynamic capacities in NGO's project management. The idea that is here presented is that PMOs in NGOs act as catalysts of dynamic capabilities enabling such organizations to achieve higher levels of performance in their projects. This argument was supported by the application of the DID method to a Brazilian NGO, in which a significant improvement in project performance was observed after the implementation of the PMO.

The practical implications of such findings involve understanding the role of the project office as a catalyst for dynamic NGO capabilities, targeting such organizations to design their project offices in order to capture, store, develop, and share the knowledge generated in each project so that the dynamic capabilities can be properly harnessed for the superior performance results of their projects.

As a theoretical contribution, this study extends the scope of the PMO as a facilitator of organizational learning through the management of projects and the development of dynamic capabilities to meet the specificities of each project and context.

However, caution should be exercised in interpreting the results, since in ex post facto studies what is generally obtained is the existence of a relationship between variables, without being able to ensure the cause-effect relationship (McMillan and Schumacher 2006). On the other hand, the results suggest relationships that can be used in future studies that have access to data from a larger number of NGOs, and of a greater temporal scope, that have implanted the PMO in their organizational structures.

Further research can also be performed using other performance metrics such as beneficiary or donor satisfaction. In addition, it would be pertinent to examine the possible differences between NGOs of different sizes, sectors of activity, and regions, whose culture and economic conjecture are different.

Appendix

Outputs – Software R

One-way ANOVA

Table Appendix – ANOVA

	Df	Sum Sq	Mean Sq	F value	Pr (>F)
Groups_Before_dummy	1	0.00125	0.00125	0.806	0.382
Residuals	16	0.02480	0.00155		

Difference-in-Differences – Multiple Linear Regression

Table Appendix – DID

Residuals				
Min	1Q	Median	3Q	Max
−0.06000	−0.04750	0.01833	0.02667	0.05000

Coefficients				
	Estimate	Std. error	t value	Pr (>ltl)
(Intercept)	0.80667	0.01382	58.373	< 2e-16 ***
Time_dummy	−0.03333	0.01954	−1.706	0.0978 .
Groups_dummy	−0.01667	0.01954	−0.853	0.4001
Mod	0.17333	0.02764	6.271	4.96e–07 ***

Signif. codes: 0 '***' 0.001 '**' 0.01 '*' 0.05 '.' 0.1 ' ' 1
Residual standard error: 0.04146 on 32 degrees of freedom
Multiple R-squared: 0.714; adjusted R-squared, 0.6872
F-statistic: 26.63 on 3 and 32 DF; *p*-value, 7.916e-09

References

Alexandrova, M., Stankova, L., & Gelemenov, A. (2015). The role of project office for project portfolio management. *Economic Alternatives, 1*, 19–30.

Alvarez, S. (2009). Beyond NGO-ization? Reflections from Latin America. *Development, 52*(2), 175–184.

Ambiental, A. G. (2015). *Perfil das ONGs ambientais* (pp. 213–252). São Paulo: Análise editorial.

Ashenfelter, O., & Card, D. (1985). Using the longitudinal structure of earnings to estimate the effect of training programs. *The Review of Economics and Statistics, 67*(4), 648–660.

Aubry, M., Müller, R., Hobbs, B., & Blomquist, T. (2010). Project management offices in transition. *International Journal of Project Management, 28*(8), 766–778.

Augier, M., & Teece, D. J. (2008). Strategy as evolution with design: The foundations of dynamic capabilities and the role of managers in the economic system. *Organization Studies, 29*(8/9), 1187.

Barney, J. B. (1986). Organizational culture: Can it be a source of sustained competitive advantage? *The Academy of Management Review, 11*(3), 656–665.

Bates, W. (1998). Improving project management. *IIE Solutions, 30*(10), 42–43.

Biesenthal, C., Vignehsa, K., Sankaran, S., Pitsis, T., Gudergan, S., Clegg, C., & Anandakumar, A. (2012). Tools of change: Exploring the duality of dynamic capabilities in project-based organizations. European group for organizational studies. Retrieved from https://opus.lib.uts.edu.au/bitstream/10453/30367/1/EGOS_Chris_Submission_V5_010612_final.doc

Block, T. R., & Frame, J. D. (1998). *The project office: A key to managing projects effectively*. Menlo Park: Crisp Publications.

Campbell, D. T., & Stanley, J. C. (1979). *Delineamentos experimentais e quase experimentais de pesquisa*. São Paulo: EPU.

Chapin, F. S. (1947). *Experimental designs in sociological research*. New York: Harper.

Cooper, R. G., Edgett, S. J., & Kleinschmidt, E. J. (1997). Portfolio management for new products. *Research-Technology Management, 40*(5), 16–28.

Crossan, M. M., Lane, H. W., & White, R. E. (1999). An organizational learning framework: From intuition to institution. *Academy of Management Review, 24*(3), 522–537.

Dai, C. X., & Wells, W. G. (2004). An exploration of project management office features and their relationship to project performance. *International Journal of Project Management, 22*(7), 523–532.

Desouza, K. C., & Evaristo, J. R. (2006). Project management offices: A case of knowledge-based archetypes. *International Journal of Information Management, 26*(5), 414–423.

Diallo, A., & Thuillier, D. (2004). The success dimensions of international development projects: The perceptions of African project coordinators. *International Journal of Project Management, 22*(1), 19–31.

Gazeta online. (2016, october 1). *Mil nascentes já estão protegidas no rio Doce*. Retrieved from http://www.gazetaonline.com.br/noticias/cidades/2016/09/mil-nascentes-ja-estao-protegidas-no-rio-doce-1013981825.html

Glaeser, E. L. (2002). The governance of not-for-profit firms. *Working paper n. 8921*. National Bureau of Economic Research, Cambridge. Retrieved from http://www.nber.org/papers/w8921

Golini, R., Kalchschmidt, M., & Landoni, P. (2014). Adoption of project management practices: The impact on international development projects of non-governmental organizations. *International Journal of Project Management, 33*(3), 650–663.

Hobbs, B., & Aubry, M. (2008). An empirically grounded search for typology of project management offices. *Project Management Journal, 39*(1), 69–82.

Hurt, M., & Thomas, J. L. (2009). Building value through sustainable project management offices. *Project Management Journal, 40*(1), 55–72.

Jalal, M. P., & Koosha, S. M. (2015). Identifying organizational variables affecting project management of fi ce characteristics and analyzing their correlations in the Iranian project-ori-

ented organizations of the construction industry. *International Journal of Project Management, 33*(2), 458–466.
Joslin, R., & Müller, R. (2015). Relationships between a project management methodology and project success in different project governance contexts. *International Journal of Project Management, 33*(6), 1377–1392.
Julian, J. (2008). How project management office leaders facilitate cross-project learning and continuous improvement. *Project Management Journal, 39*(3), 43–58.
Kwak, Y. H., & Dai, C. (2000). Assessing the value of project management offices (PMO). PMI Research Conference, Paris, France. Retrieved from https://www.pmi.org/learning/library/assessing-value-project-management-offices-8527
Killen, C. P., Hunt, R. A., & Kleinschmidt, E. J. (2008). Learning investments and organizational capabilities. *International Journal of Managing Projects in Business, 1*(3), 334–351.
Lacruz, A. J., & Cunha, E. A. (ahead of print). *Escritório de projetos em entidades do terceiro setor: um estudo ex post facto*. REAd - Revista Eletrônica de Administração.
Lacruz, A. J. (2014). *Gestão de projetos no terceiro setor*. Rio de Janeiro: Elsevier.
Lacruz, A. J. (2015). Project management office in nonprofit firms: An application in NGO Instituto Terra in Brazil. *Business and Management Review, 5*(4), 510–523.
Leitão, M. (2016, april 19). *Esperança nos olhos d'água*. Jornal O Globo. Retrieved from https://oglobo.globo.com/sociedade/sustentabilidade/instituto-inicia-projeto-para-proteger-nascentes-do-vale-do-rio-doce-1-15922872
Liu, S. (2015). Effects of control on the performance of information systems projects: The moderating role of complexity risk. *Journal of Operations Management, 36*, 46–62.
McKelvie, A., & Davidsson, P. (2009). From resource base to dynamic capabilities: An investigation of new firms. *British Journal of Management, 20*(Suppl.1), S63–S80.
McMillan, J. H., & Schumacher, S. (2006). *Research in education: Evidence-based inquiry*. Boston: Pearson.
Nelson, R., & Winter, S. (1982). *An evolutionary theory of economic change*. Cambridge: Harvard University Press.
Pellegrinelli, S., & Garagna, L. (2009). Towards a conceptualisation of PMO's as agents and subjects of change and renewal. *International Journal of Project Management, 27*(7), 649–656.
Penrose, E. (1959). *The theory of the growth of the firm*. London: Basil Blackwell.
Petit, I., & Hobbs, B. (2010). Project portfolios in dynamic environments: Sources of uncertainty and sensing mechanisms. *Project Management Journal, 41*(4), 46–58.
Project Management Institute. (2013). *Project Management body of knowledge: A guide to the Project Management body of knowledge*. Pennsylvania: PMI.
R Core Team. (2017). *R*: A language and environment for statistical computing (version 3.3.3). [Software]. R foundation for statistical computing, Vienna, Austria. Retrieved from https://cran.r-project.org/src/base/R-3/R-3.3.3.tar.gz
Silva, C. E. G. (2010). Gestão, legislação e fontes de recursos no terceiro setor brasileiro: uma perspectiva histórica. *Revista de Administração Pública, 46*(6), 1301–1325.
Schumpeter, J. A. (1934). *Theory of economic development*. Massachusetts: Harvard University Press.
Schumpeter, J. A. (1942). *Capitalism, socialism, and democracy*. New York: Harper.
Stanleigh, M. (2006). From crisis to control: New standards for project management. *Ivey Business Journal, 70*(4), 1–4.
Teece, D. J., Pisano, G., & Shuen, A. (1997). Dynamic capabilities and strategic management. *Strategic Management Journal, 18*(7), 509–533.
Teece, D. J. (2007). Explicating dynamic capabilities: The nature and microfoundations of (sustainable) enterprise performance. *Strategic Management Journal, 28*(13), 1319–1350.
Teece, D. J. (2009). *Dynamic capabilities & strategic management: Organizing for innovation and growth*. New York: Oxford University Press.
Tenório, F. C. (1999). Um espectro ronda o terceiro setor: o espectro do mercado. *Revista de Administração Pública, 33*(5), 85–102.

Turner, J. R., & Muller, R. (2003). On the nature of the project as temporary organization. *International Journal of Project Management, 21*(1), 1–8.

Unger, B. N., Gemünden, H. G., & Aubry, M. (2012). The three roles of a project portfolio management office: Their impact on portfolio management execution and success. *International Journal of Project Management, 30*(5), 608–620.

Winter, S. G. (2003). Understanding dynamic capabilities. *Strategic Management Journal, 24*(10), 991–995.

Williamson, O. E. (1975). *Markets and hierarchies*. New York: Free Press.

Williamson, O. E. (1985). *The economic institutions of capitalism*. New York: Free Press.

Yazici, H. J. (2009). The role of project management maturity and organizational culture in perceived performance. *Project Management Journal, 40*(3), 14–33.

Zollo, M., & Winter, S. G. (2002). Deliberate learning and the evolution of dynamic capabilities. *Organization Science, 13*(3), 339–351.

Chapter 4
Sustainable Development of Small and Medium-Sized Enterprises in Disadvantaged Regions: Impact of Knowledge and Innovation

Mário Franco and Heiko Haase

4.1 Introduction

Knowledge and innovation are critical resources through which small and medium-sized enterprises (SMEs) create new products or invest in existing resources with a greater potential to create wealth (Acs and Armington 2006; Carlino and Kerr 2015). Therefore, firms should concentrate on these phenomena to reinvent their business and examine new opportunities to make them more competitive (Kamalian et al. 2011). In order to achieve sustainable competitive advantages, leveraging knowledge becomes crucial for SMEs (Pillania 2006). Moreover, O'Gorman (2001) and Moore and Manring (2009) showed that entrepreneurs' strategic choices influence SMEs' growth and sustainable development. In this context, SMEs in disadvantaged regions must make an additional effort, as change is slightly slower, and access to innovation and knowledge presents greater difficulties.

Most studies on innovation and knowledge management are linked to economic development. However, these studies focus on "how" and "why" innovation and knowledge are developed (Bruton et al. 2008), with few studies (e.g. Naudé 2013) showing the impact of these phenomena on the development of disadvantaged regions and SMEs' sustainable development. Nevertheless, Acs and Armington (2006) concluded that not only firms' size is important but also that the presence of innovative entrepreneurs is important for regional development, leading to increased

M. Franco (✉)
Management and Economics Department, CEFAGE-UBI Research Center, University of Beira Interior, Covilhã, Portugal
e-mail: mfranco@ubi.pt

H. Haase
Department of Business Administration, Center for Innovation and Entrepreneurship, University of Applied Sciences Jena, Jena, Germany
e-mail: heiko.haase@eah-jena.de

M. Peris-Ortiz et al. (eds.), *Knowledge, Innovation and Sustainable Development in Organizations*, Innovation, Technology, and Knowledge Management,
https://doi.org/10.1007/978-3-319-74881-8_4

employment. Praag and Versloot (2007) also contributed to understanding of the entrepreneur in economic development, job creation, increased productivity and innovation. Nevertheless, how SMEs can adopt knowledge and innovation for the development of disadvantaged regions remains underexplored.

Given these gaps identified in the literature, this study intends to investigate the influence of knowledge and innovation on SMEs' sustainable development, thereby stimulating the growth and development of disadvantaged regions. To attain these objectives, the research questions are as follows: (1) How do SMEs in disadvantaged regions innovate? (2) How do knowledge and innovation in SMEs allow their sustainability and the development of the regions they belong to? In order to answer these questions, a qualitative study in six SMEs in the Beira Interior Region in Portugal was conducted. With this approach, this research contributes to better understanding of the role of knowledge and innovation as tools for the development of SMEs in disadvantaged regions. The intention is also to identify the relevance of knowledge and innovation as sources of regional development.

The structure of this chapter is as follows: Sect. 4.2 provides a literature review about the role of knowledge and innovation in economic development in general and for SMEs in disadvantaged regions in particular. Afterwards, Sect. 4.3 refers to the methods used in this qualitative study. Section 4.4 characterises the selected cases, and Sect. 4.5 provides a cross-case analysis. In Sect. 4.6, the chapter concludes with some final considerations and implications for theory and practice.

4.2 Literature Review

Knowledge allows firms to share resources and enhance their innovative capacities. Nevertheless, certain doubts still remain concerning knowledge and innovation for SMEs' economic development, despite the increasing attention paid to this topic. Therefore, in this section, some of the underlying concepts are discussed along with the main reasons leading SMEs to adopt knowledge and innovation in disadvantaged regions.

4.2.1 The Role of Knowledge and Innovation in Economic Development

Various studies confirm the connections between innovation, entrepreneurship and economic growth. These studies show clearly positive results, namely, in relation to the product, productivity and employment (Audretsch and Fritsch 2002; Acs and Armington 2004; Audretsch and Keilbach 2004; Fritsch and Müller 2007). The entrepreneur can contribute in various ways: in products, innovation, new procedures, supply sources, exploring new markets and work or management methodologies (Szirmai et al. 2011). However, not all entrepreneurs are innovative or allow development, as although jobs are created many are poorly paid (Stam and Wennberg 2009).

According to Audretsch (2003), some studies also confirm a positive relationship between knowledge and innovation and economic development. Through activities related to innovation, firms have a positive impact on development (van Stel et al. 2005). However, to ensure this, there must be changes in terms of structure, policies, processes and production methods, as well as for efficient use of existing resources (McMillan and Rodrik 2011). With these changes, firms can contribute by facilitating the introduction of available resources, increasing their efficiency (Acs and Storey 2004), lowering intrinsic costs and supporting economies' structural changes (Gries and Naudé 2010).

With conditions adjusted to the market and development, more firms or investors will be attracted to the region or country, to carry out their projects. Consequently, this will provide established firms with growth opportunities through an increased number of collaborators, clients and suppliers in the market. According to Schwab (2014), the existence of financial protection allows individuals to make riskier decisions. The return will benefit the economy due to job creation and innovative ideas that are the result of particular firms and industries or social and demographic trends (Drucker 1985), consequently adding to economic growth.

It is also argued by Acs (2008) that knowledge and innovation's influence on the economy can benefit firms, with the latter having a positive effect on development. In this connection, Audretsch and Keilbach (2004) formulate the hypothesis that economic growth results from innovative and knowledge activities and that the greater those activities, the higher the level of economic growth. So knowledge and innovation have an important role in economic growth and the respective development (Baumol et al. 2007) favouring new forms of firm organisation (Kim et al. 2006).

4.2.2 The Role of Knowledge and Innovation for SMEs in Disadvantaged Regions

Location as a factor of firms' and also regions' growth and development was considered important for companies and entrepreneurs (Porter 2000). The sources of growth became more diverse as the population grew, confirming the possibility of firms having advantages in terms of infrastructure, values, culture and institutions (Porter 1990). With these new possibilities for growth offered by location, firms have three choices (Chandler 1962): (i) continue to grow alongside the expanding population, (ii) seek to conquer new markets and (iii) develop new products and increase their supply capacity.

One of the advantages of choosing an appropriate location is the concentration in the same place of firms with similar characteristics that convey information. This topic is explored extensively by Porter (1990, 2000), describing the important role of these business concentrations. According to this author, these business nuclei bring firms close to clients, suppliers of specialised products and services, firms in the same sector and universities, institutions and associations. This relationship facilitates improved performance and growth through cooperation, with the share of knowledge about technology and markets, raising firms' competitiveness.

In this context, knowledge management and innovation become important phenomena in regional, national and international development, changing traditional economies into new economies. Attention was also drawn to SMEs (Gupta et al. 2013), which have a central role in regional dynamics and frequently account for a great proportion of total jobs created. Several studies (e.g. Carree and Thurik 2003; van Stel et al. 2005; Wennekers et al. 2005) have dealt with the impact of firm creation and SMEs on economic growth, by stimulating the economy, creating jobs and integrating the unemployed and people at a disadvantage, strengthening social cohesion in less favoured regions. The existence of firms stimulating the knowledge base and knowledge spillovers in regions is one of the main drivers of employment and regional development (Asheim et al. 2011).

Differences exist between the rates of entrepreneurship in richer and poorer countries, which are also reflected in the richest and poorest regions of each country. This situation is due to the fact that in richer countries or regions, the entrepreneur looks for opportunity, whereas in less favoured countries or regions, the entrepreneur acts through need (Wennekers et al. 2005). Indeed, innovation and knowledge differ according to each country or region's economic development (Szirmai et al. 2011). While in more developed regions, innovation takes place through the formulation and production of products and services that are new to the firm or market, in less favoured regions innovation is understood as the use of products, methods or services to the benefit of firms or the economy itself.

Various methods are observed regarding regional growth, namely, the population's income or GDP per capita, the number of employees and amount of production. Acs and Armington (2006) studied economic growth at the geographical level, specifically in metropolitan areas, bearing in mind the interrelation between entrepreneurship and growth in the rate of employability. In carrying out their research, they concluded that entrepreneurship has a positive influence on new firm creation and that the first years of activity are the most important for increased employability when these firms are successful.

In many cases, entrepreneurship and innovation are encouraged by the State or by regions, through tax incentives or subsidies, something which, according to Florida (2005), on its own is not enough to attract the necessary people whose talent and creativity are central to the region's development. This author argues that the response to this situation comes from regions' creativity in order to attract entrepreneurs, although certain conditions must be in place. Also according to Florida (2005), this creativity can be understood as being of three types: (a) creativity of technology or innovation, (b) economic or entrepreneurial creativity and (c) artistic or cultural creativity. These three types of creativity are interrelated, so that for a region to develop innovation and knowledge activities, it has to create the right conditions for innovation, art and culture. Regions must attract different types of creative people, with a variety of mentalities and receptiveness to ideas (Florida 2003).

Therefore, inhabitants, local authorities, the State, universities and firms have a fundamental role (Maillat 1992). Regions with these conditions attract people who are more focused on creativity, with different ideas and projects, leading to increased innovation, job creation, business and technology, enabling the region to increase its economic development (Florida 2003). Porter (1990) highlighted that it is in that innovation, both in terms of products and how to produce them, that each country or region's competitiveness is found. For Porter (1990), this is the only way to maintain competitiveness, beginning in a procedure with a local character.

In this line of thought, from a perspective of joining SMEs and the people living in a region, Dijkstra et al. (2011) emphasised that regional competitiveness is formed of the capacity to attract and create an environment where firms and citizens will wish to set up in a region and create new jobs, increasing the presence of existing firms in the market. Indeed, regions have an important role in promoting the search for entrepreneurs to set up in certain areas, to develop and expand their activities. At the regional level, the role of SMEs as promoters of commercial dynamics is centred on their innovative skills and knowledge, as the key to regional development and sustainability. SMEs contribute to improving the population's quality of life, by providing good conditions for future growth, and also supporting large companies (OECD 2008).

4.3 Methods

4.3.1 Type of Study

According to the aims of this study, the qualitative method of an exploratory nature was chosen. This type of qualitative and exploratory study allows clarification and understanding of a problem when there is an absence or uncertainty regarding the nature of knowledge. So as the aim was to research a little studied subject, this was the most appropriate method for the purpose. According to Yin (2013), this type of study allows the researcher to select the elements available and therefore obtain a sample of what can be the universe of the research. Qualitative research also attempts to understand the samples through observation, description and interpretation of the data obtained.

4.3.2 Case Selection

Within qualitative research, the case study method was chosen, as this fulfils the objective of detailed analysis of a unit, environment or simply a subject or event (Yin 2013). In particular, we selected six SMEs (cases) in the Beira Interior Region of Portugal in order to scrutinise the role of knowledge and innovation as mechanisms for SMEs' sustainability in disadvantaged regions. These firms fulfil the SME criterion established by the European Commission (2003), having between 10 and 250 employees and an annual turnover no greater than 50 million euros and an annual balance sheet no greater than 43 million euros.

Initially, firms were contacted by telephone in order to present the study in question, describing the objectives, the need for their collaboration and the contribution hoped for. Next, the firms that agreed to collaborate were contacted to find out their availability to carry out the study, namely, with the founder, administrator or manager. Table 4.1 presents a brief characterisation of the six SMEs selected.

Table 4.1 Characterisation of the firms interviewed

Cases/firms	CACE	Headquarters	Year of foundation	Employees	Operating in
1	Milk and dairy products industry	Vila Velha de Rodão	2003	10	National and international market
2	23,690 – Manufacture of other concrete, plaster and cement products	Proença-a-Nova	1983	16	National and international market
3	28,250 – Manufacture of nondomestic equipment for refrigeration and ventilation	Castelo Branco	1977	50	National and international market
4	46,390 – Sale and distribution of food products	Castelo Branco	1982	14	Regional market
5	22,292 – Manufacture of display cabinets	Tortosendo	1988	75	National and international market
6	46,341 – Wholesale of alcoholic and non-alcoholic drinks	Guarda	2000	10	National and international market

4.3.3 Data Collection and Analysis

According to Eisenhardt (1989), data collection includes the following stages: (1) documentary collection and its analysis, (2) surveys, (3) interviews, (4) studies through observation and (5) data analysis. In this study, information was gathered through documentary analysis and interviews held in the SMEs willing to collaborate. This was carried out aiming to obtain the data necessary to support the empirical analysis.

The field work included: (a) visits to the firms that agreed to collaborate, with the opportunity to confirm personally the entrepreneurial and innovative initiatives, and (b) holding interviews with those in charge of the firms or designated to answer the questions drawn up. Interviews are frequently used for this type of study, as they allow greater understanding of the issues and provide vital information to develop the research and deepen the subjects studied. This type of method also presents advantages such as flexibility, making it easier for the interviewees to explain their answers (Yin 2013).

The interview script was semi-structured, as in this model topics tend to be more specific with a combination of open and closed questions. This type of interview had the advantage of obtaining more efficient and detailed information, favouring the emergence of spontaneous answers provoked by the interaction between interviewee and interviewer. The interviews were held in May 2015 and lasted between 30 and 60 min. Table 4.2 presents a brief characterisation of the interviewees.

Table 4.2 Characterisation of the officials interviewed

Cases/ firms	Gender	Interviewee's function	Age	Qualifications	Date of interview	Length of interview
1	Female	Owner-manager	43	Diploma	07/5/2015	50 min
2	Male	Production manager	33	Diploma	08/5/2015	40 min
3	Male	Owner-manager	62	7th year of school (former system)	12/5/2015	25 min
4	Male	Sales director	48	Degree	12/5/2015	1 h 20 min
5	Male	Head of sales/ financial director	41	Degree	13/5/2015	45 min
6	Male	Manager	46	11th year of school	14/5/2015	60 min

After carrying out the interviews, a report was drawn up with the information obtained. The reports of each firm were sent to the interviewees afterwards for their appraisal and approval. After data collection, resorting to the techniques described above, the data was subject to analysis and treatment.

4.4 Case Studies

In this section, based on the two units of analysis ((i) the role of knowledge and innovation in the firm and (ii) business, economic and regional development), each of these units in each of the six cases studied will be described.

4.4.1 Case 1

Role of Knowledge and Innovation in the Firm

This firm has been involved in industrial production, namely, cheese production, since 2003. Located in the district of Castelo Branco (Portugal), this SME has ten employees. It operates in national markets but also has projects to expand sales internationally. The role of knowledge and innovation is present in the firm. According to the owner-manager, its innovation strategies have included "some partnerships with entities that can develop new products. Internally, we adapted our existing products, with a change in taste, towards a more gourmet segment, also improving the aesthetics and packaging".

The interviewee highlights the importance of knowledge for innovation, stating that "we have achieved better quality in our products, which is our main focus, and this is reflected in the awards won, invitations to participate in events or from Castelo Branco Local Authority to move our production to Castelo Branco". The firm's investments have focused more on the product level, as stated by the interviewee. Only at the beginning of the activity did they make major investment in semi-automatic equipment, one of the first in the region. Around 2 years ago, the firm also invested in chambers to mature and dry the cheese. Annually, investment has been about 30,000 thousand euros, with a return between 1% and 5%.

The interviewee adds that although knowledge and innovation are a constant parameter in the firm's focus, some difficulties are faced, specifically: "the financial aspect. Given the financial crisis, we have been growing, but not as much as we would like. The margins are becoming smaller, as some raw material increased in price. Also the bureaucratic aspects, especially in the food sector, every day we have documents to be filled in, and some of this is done manually".

Business, Economic and Regional Development

The firm has grown in recent years despite the crisis in the market slowing this down. For the interviewee, the main reason for that growth and firm performance is essentially "product quality". This is apparent from the beginning of the production process "from raw material selection to the finished product, we look for quality", allowing them to stand out from the competition in this aspect. This performance and its improvement is possible through knowledge and innovation, as "the fact of focusing on other markets and introducing new products will have an impact internally as we'll have to increase production, increase our production capacity and take on more workers", the interviewee says. This improvement in the firm led to it being declared a "Leading SME" in 2014.

Regarding economic development, the interviewee also mentions that "if we grow, it's good for the people who are with us, as we can take on new people, create better conditions for those already working in the company". With this growth, firms can help the region regarding its development, new job creation and wealth generation. This growth "brings more people to the regions, and they will take on other projects and activities, with new ideas", the owner-manager also states. Firms already in the region, if they are in "good health", by generating growth and employment will attract new people, who, according to the interviewee, "also set up other companies, exploring the tourism and restaurant sectors, allowing development", but also through "attracting young people" to inland regions with an ageing population.

4.4.2 Case 2

Role of Knowledge and Innovation in the Firm

This firm has operated in the construction and industrial production sector since 1983. It is located in the Industrial Estate in Proença-a-Nova (Portugal) and currently has 16 employees. The firm operates in both the national and international markets, and "always tries to be innovative", placing a greater emphasis on construction methods. According to the interviewee, the company "concentrates on product differentiation, but also on expanding its market".

Although the firm has concentrated on national or international fairs to become known, it is currently focusing on a more face-to-face strategy with regard to its clients. The firm considers knowledge and innovation as the way forward, as according to the interviewee, "by creating innovation, the aim is to increase turnover, open up new markets and create motivation inside the firm for something new, and new ideas can be created by the management or one of the employees". Concerning innovation, the company recently adopted a new building method called four squared and a new covering method, Ecofinish. This type of investment represents amounts between 30,000 and 100,000 euros, giving a return between 11% and 15%.

For the interviewee, the fact the firm is situated in an economically depressed area has its advantages and disadvantages. He identified communication and transport costs as a disadvantage. However, "the cost of living is lower", which is reflected in the employees. One difficulty identified by the interviewee is "bureaucracy". He says, "some projects we develop, due to the bureaucracy, end up not moving forward and there's a loss of motivation at certain times between finishing the project and passing to the construction stage".

Business, Economic and Regional Development

At the firm level, the interviewee considers consultation with clients as an indicator of performance, as at the moment they do not advertise generally. The interviewee says that knowledge and innovation allow "an increase in the firm's wealth, the creation of new markets, more collaborators and clients". Consequently, it means "always being different from the competition" through "innovation, with products that are ours alone, exclusively".

At the regional level, the interviewee mentions that "firms' role is principally social". For the interviewee, "regionally, with our main objective being increased sales, we will have more collaborators, increasing the region's population. Some weren't from the area, they moved here, and logically that has an impact on the region as new business develops and other firms also look more to the region".

4.4.3 Case 3

Role of Knowledge and Innovation in the Firm

The firm manufactures nondomestic equipment for refrigeration and ventilation. It is located in the town of Castelo Branco (Portugal) and has 50 employees. This firm is present in the national and international market. Its knowledge and innovation strategy includes looking for new markets beyond the regional or national ones. They concentrate on the internationalisation of their products and services.

According to the interviewee, that attitude means "keeping up to date with the market", and so they focus on innovation in terms of decorations, product image and new products, representing an annual investment of between 30,001 and 100,000 euros. The results of these innovations were, "for example, these years of crisis not affecting the firm, with a growth in sales", with a return between 6% and 10%.

Since the firm focuses mainly on the international market, according to the interviewee, the greatest difficulty is in relation to suppliers being able to respond to the firm's needs. He also says that "locally, we can't find firms with the capacity to respond to our needs, for example, for supplies or services, and so we have to go to firms in the Lisbon area or abroad, for example, to Denmark and Italy".

Business, Economic and Regional Development

Concerning the firm's performance, the interviewee identifies "the demand from the international market showing an increase since the peak of the crisis due to our expansion into new markets" as an indicator of improved company performance. He also states that its entrepreneurial capacity has an impact on its performance, as "the international market is very demanding, and so we need higher quality to be able to beat the competition in those markets". In order to match or rise above that competition, above all internationally, it was "necessary to make some investment, constantly, to modernise the company. Concentrating on quality and being careful not to make mistakes and get a bad name in the market".

As for economic and regional growth, the interviewee says he was not the right person to answer, as for him, entrepreneurship and innovation only creates jobs. The manager also believes that entrepreneurship and innovation in SMEs do not allow regions to develop, explaining that "a firm like ours, which gets raw material outside, produces here and then takes it away again, does not create regional wealth. Manufacturing firms prefer to remain in cheaper regions such as the coastal strip, where costs are lower".

4.4.4 Case 4

Role of Knowledge and Innovation in the Firm

This firm is located in the district of Castelo Branco and has 14 collaborators. It is an SME selling frozen food products. Knowledge and innovation in the firm are constant factors emphasised by management. Some strategies such as the internal organisation of operating processes and methods were adopted, namely, "some technology so as to provide the sales-people with better information, allowing them to show the products more quickly, and also orders. It also lets us know about clients' purchases in detail". According to the interviewee, this process also allowed optimisation and control of costs.

Although the interviewee agrees they are situated in a less-favourable region, it is necessary to manage costs well and focus on systems that can lead the company to efficiency. For the interviewee, this entrepreneurial and innovative attitude "accelerates work processes and firms can create more jobs", with this being reflected in the "net result and a rise in the value of sales in these last years". The greatest obstacle to the firm's growth, as well as the focus on knowledge management and innovation, is linked to "bureaucracy and justice". He also recognises that in a less favoured region "there are fewer resources, in terms of information and training".

Business, Economic and Regional Development

As for business development, the interviewee identifies quality in service provision. Through that quality created by entrepreneurial and innovative skill, the main result concerns sustainability. Quality also promotes "client loyalty". According to the interviewee, "if we are not entrepreneurial, we become maladjusted and unable to maintain the firm's sustainability". For him, the relationship between entrepreneurship and economic development is linked in all sectors, as this allows greater mutual dynamics. This aspect in firms provides a "demand for services in other companies, creating wealth".

For the sales director, "it is firms that stimulate the region, through job creation. They also support regions' development, participating with universities and schools, allowing these to collaborate with firms in small tasks such as marketing or creativity". Finally, the interviewee says, "there's no doubt that companies allow regions to grow. However, there should be more help for investment in inland regions".

4.4.5 Case 5

Role of Knowledge and Innovation in the Firm

The firm is located in the Industrial Estate of Covilhã (Portugal) and operates in industrial production. It was founded in 1988 and currently has 75 employees. Its markets are national and international, with the latter being more important. The firm is seen to be active, seeking new markets and development opportunities.

Speaking about subjects related to knowledge and innovation, the interviewee says this is a firm that always accompanies market needs. This accompaniment includes "development of new products (we have a department for conceiving and developing new products or ideas from clients), new technology, production methods. We form new partnerships according to our needs". According to the interviewee, recently some innovations in the manufacturing process, machines and technology, and the firm's organisation in terms of communication and employee training were essential to "increase turnover and improve cost control. We managed to expand the market, capturing new clients. We improved our response capacity and the quality of our products".

Entrepreneurship can be named as an important factor for the firm's innovation, as according to the businessman, "without entrepreneurship we can't innovate". This implies the need to "analyse the market and its needs to be able to innovate". The firm's location in an economically depressed area does not prevent it from being entrepreneurial and innovative, the interviewee says. To do so, the firm's strategy includes a "presence in international fairs in the sector, to accompany market trends. Visits to other countries to analyse markets, see new ideas and new products". The greatest difficulties faced by the company in being entrepreneurial and innovative are related to "timing". In the current market, "clients want the product for yesterday", or "make changes", which have to be adopted by the firm.

Business, Economic and Regional Development

In the interviewee's opinion, business development is reflected in "the firm's net results". However, he adds "opening up new markets", which in turn will increase the number of clients, and "we do not depend on just one market". The businessman adds this point stating that the firm's entrepreneurship and innovation increased the "firm's dynamics". This allows it to compete both nationally and internationally. Although the latter is more "aggressive", the market is "vast, which allows clients to have more than one supplier".

When turning to matters related to economic and regional development, the interviewee has no doubt that entrepreneurship and innovation allow both companies and regions to grow. He says, "value is created through entrepreneurship. Jobs are created, which is important socially. With more projects and work, we can take on more workers". He adds that the role of firms in regions includes the creation of "employment, increasing purchasing power, and therefore increases the money in circulation. Without local companies, the economy stagnates, people have to leave the regions". This is possible through "entrepreneurship and innovation, so that firms can grow and consequently influence the region's growth".

4.4.6 Case 6

Role of Knowledge and Innovation in the Firm

The firm's business is the sale of drinks and it is situated in Guarda (Portugal). It started its activity in 2000 and now has ten employees. It operates in the national, particularly regional, and international market. To allow the firm to grow in the market, it has adopted some strategies of knowledge and innovation. According to the interviewee, these have included a "new computing system, new products and market expansion, outside Portugal, more specifically in Spain. Also investment in larger premises so as to be able to rent space to other firms wanting to begin operations".

He recognises, however, that being entrepreneurial and innovating in economically disadvantaged regions is not easy and the most important thing is to "have a good relationship with the client" and provide some "new products, as clients often expect novelties". So as in the previous cases, this manager identified the net result or sales as the main effect of entrepreneurship and innovation in the firm. Annual investment is around 30,000 euros reflecting an approximate return of 2%. With the new computing system introduced, the firm is able to "have more detailed information about clients, consumption, variations and other data".

When asked about the difficulties encountered regarding knowledge and innovation, the businessman says that "all the investment we make is with our own capital. Bank access is more limited. Many firms find it difficult to make projects, to get funding to be able to invest. We also have some internal difficulty in adapting to change".

Business, Economic and Regional Development

One of the points the manager recognises as an indicator of business development is related to meeting the objectives established each month. He adds that entrepreneurship and innovation lets clients have "greater confidence in our services" and that in relation to the competition, clients "perceive better service". The businessman believes that entrepreneurship and innovation support economic development, as this "creates more employment, which generates commerce and the money in circulation" especially in less favoured regions due to the influence of "increased employment".

According to the interviewee, firms also play a social role of supporting associations and organisations; "if firms are in a good position, they can help the region, causing other organisations to develop, promoting the good name of the company". He also mentions that entrepreneurship and innovation in SMEs allows regions to grow and develop, as "if firms have growth and work, this also gives the region a name. They can innovate, have other services, which reflects the region's image in relation to others, increasing other firms' demand for the region's services and products".

4.5 Cross-case Analysis

4.5.1 The Role of Knowledge and Innovation in the SMEs Studied

The trends towards knowledge and innovation are constant factors in the firms/cases studied here. These are tools for firms' growth and development in order to consolidate their position in the market. This idea is also defended by Audretsch and

Keilbach (2004). Although the SMEs studied here are situated in economically disadvantaged areas, with financial, human and tangible resources relatively lower (Saxena 2012) than firms located on the Portuguese coast, it is by choosing an entrepreneurial and innovative route that they manage to retain their competitiveness (OECD 2008).

Therefore, the role of knowledge and innovation includes a strategy that the firms analysed adopt as a way to overcome the current economic situation felt in Portugal. For these companies/study cases, choosing knowledge and innovation is important in aiming for growth and sustainability. "Without entrepreneurship, we can't innovate" (Case 5) and "if we are not entrepreneurs, we become maladjusted and unable to maintain the firm's sustainability" (Case 4), and also for Case 2, entrepreneurship should "always be focused on innovation".

According to Case 6, this path "creates more employment, which generates commerce and the money in circulation", through "concentrating on other markets and the introduction of new products, which has an impact internally as we'll have to increase production, increase our manufacturing capacity and take on more workers" (Case 1). Also Case 2 defends knowledge and innovation, as "by creating innovation, the aim is to increase turnover, open up new markets and create motivation inside the company for something new, and new ideas can by created by management or by any employee".

For Case 5, "value is created through entrepreneurship. Jobs are created, which is important socially. With more projects and work, we can take on more workers" which is also reflected in the firms' results. As referred to in Case 1, "we have achieved better product quality, our main focus, which is reflected in the awards won". Knowledge and innovation in the firms/cases represent an important mechanism with a view to improving their capacities to "always be different from the competition" (Case 2) as also mentioned by Porter (1990), especially as "international markets are very demanding, so we need higher quality to be able to rise above the competition in those markets" (Case 3).

This path via knowledge and innovation gives firms continuous development of products, production processes, new technology and markets (Szirmai et al. 2011). In Case 5, the "development of new products (we have a department for conceiving and developing new products or clients' ideas), new technology, production methods", and "we manage to expand the market, capturing new clients. We improve our response capacity and the quality of our products". This result can be observed, as mentioned by Case 3, that, "for example, these years of crisis have not affected the company, with a growth in sales". Various authors mention the importance of knowledge and innovation as a competence of firms and entrepreneurial agents for the differentiation of products, markets or processes (e.g. Carlino and Kerr 2015). That focus by firms is related to their growth and development.

For firms to ensure clients' satisfaction, the search for knowledge, entrepreneurship and innovation has been a constant concern (Kamalian et al. 2011). Related to this aspect, Case 6 says its clients have "greater confidence in its services" and that compared to the competition, clients "perceive better service". Case 4 also mentions that through this characteristic, quality is increased, leading to "client loyalty".

Referring to Omar et al. (2009), some difficulties recognised for the development of knowledge and innovation are, for example, access to finance, bureaucracy, obtaining a specialised workforce, time to respond to clients or difficulties in making structural changes.

Concerning finance, firms (Cases 1 and 6) indicate that "given the financial crisis, we have been growing, but not as much as we'd like. Margins are becoming smaller, as some raw material has risen in price" (Case 1) and that "often as there's limited access to banks, all the investment we make is with our own capital" (Case 6). Bureaucracy is also present in the firms studied, bringing some difficulties, as it delays processes or prevents firm development (Bartlett and Bukvić 2001). This can be seen in Cases 1 and 2, which in the latter means "ending up not being able to go ahead with many projects". Also in Case 4, bureaucracy becomes an obstacle for the firm's development and growth regarding entrepreneurship and innovation.

It is more difficult to obtain a workforce for SMEs, which is felt in the firms when they look for certain specialised services. This situation arises in Case 3, where "we can't find firms with the capacity to respond to our needs", and so the firm looks outside its region. According to Kofjač et al. (2010), firms face difficulties in production times in relation to clients' demands, as their production capacity may not be enough or through a lack of knowledge in the area. In Case 5 production times are found to be a difficulty for entrepreneurship and innovation, as "the clients want the product for yesterday". Also in Case 6, adaptation to knowledge in the areas or technology emerges as a barrier to entrepreneurship.

4.5.2 Knowledge and Innovation as a Means for SMEs' Sustainability

The tendency to turn to knowledge and innovation has always been present in the six firms studied. To ensure their continued competitiveness and growth, from an early stage they have focused on knowledge and innovation. That attitude influences not only the SMEs but also the regions they belong to (Dijkstra et al. 2011; OECD 2008). According to Popova and Sharpanskykh (2010), certain indicators show firms' results and their performance. Highlighted among them are the quality of products or services, jobs created, profit, market demand, productivity and new markets. For Cases 1 and 4, the main indicator is related to the "quality of the product or service" that the firm provides.

In Case 2, the interviewee mentions that the fundamental indicator of performance is related to "consultation with clients", as at the moment they do not widely publicise their products. For Case 3, the sign the firm is performing well is in "demand in the international market showing an increase since the peak of the crisis due to expanding into other markets". For Case 5, this performance is related to the "firm's net results", while for Case 6 it is associated with "meeting the objectives that are set every month".

According to Audretsch and Keilbach (2004) and Fritsch and Müller (2007), firms' sustainability is related to knowledge and innovation. This performance is attainable, according to Case 1, as "focusing on other markets and introducing new products will have an impact internally as we'll have to increase production, increase our manufacturing capacity and take on more workers". This idea is defended by McMillan and Woodruff (2002) and Kim et al. (2006). These events (Cases 2 and 4) can "increase the firm's wealth, create new markets, increase collaborators and clients" and "a demand for services in other firms, creating wealth", leading to greater firm sustainability (Carlino and Kerr 2015). This point is also seen in the international market, as according to Case 3, "the international market is very demanding, and so we need higher quality", a fact also stressed by Elberse and Eliashberg (2003).

One of the aspects described by Case 5, as an indicator of good performance, is related to the "firm's net result". In the final case (Case 6), performance is indicated by fulfilment of the "objectives that are set every month", most of which can be achieved. Many authors approach the subject of economic and regional development (e.g. Dijkstra et al. 2011; Acs and Armington 2006; Florida 2003). In the range of firms studied, all except Case 3 argue that knowledge and innovation go side by side with economic or regional development. For Case 1, "if we grow, it's good for the people that are with us and we take on new people", and with this growth firms can support the region in its growth, as this "brings more people to the regions, and they will develop other projects and activities, with new ideas".

For Case 2, "firms' role is mainly social". Also at the regional level, "with increased sales, we'll have more collaborators, increasing the region's population. Some were not from the area, they moved here and logically this has an impact on the region as they set up new businesses and other firms also look more to the region". However, regionally the firm (Case 3) brings few benefits and little growth, as "a firm like ours, that gets raw material from outside, produces here and takes it away again does not generate regional wealth". In Case 4 there is no doubt that firms allow regional growth. It is firms that stimulate the region, through job creation. They also support regions in their development, participating with universities and schools, allowing these to collaborate with firms in small undertakings.

The interviewee in Case 5 also mentions the importance of knowledge and innovation for regions in social terms, highlighting that "value is created through entrepreneurship". Case 6 also argues that knowledge and innovation develop regions as they "generate more employment, which increases purchasing power" and "without local firms, the economy stagnates, people have to leave the region". In addition, firms have a social role (Cases 2 and 5), because "if firms are healthy, they can help the region, causing other organisations to develop, giving the company a good name". Finally, Case 6 states that "if firms have growth and work, this also gives the region a name. They can innovate, have other services, which reflects the region's image in relation to others, increasing other firms' demand for regional services and products". It can therefore be stated that the firm's performance has an impact on regional economic development.

Through the mechanism of knowledge and innovation, the firms studied accept the demand for improved sustainability. This includes the conception of new products, equipment, production methods and organisational processes and searching for new markets (Carlino and Kerr 2015). In this aspect, the firms try to generate the development and growth of their organisations, and consequently their sales, the supply of products and market expansion, allowing the firm's sustainability through a more effective and efficient response. Although not the opinion of all the cases analysed, this development influences firms' role in the regions they are part of. This growth allows the creation of new jobs, improved conditions for employees and support for other organisations in the region and generates more wealth and social and professional well-being (Dijkstra et al. 2011).

The following table summarises, for each of the two units of analysis identified in this study, the most important aspects retained from the empirical evidence obtained in each of the SMEs studied (Table 4.3).

4.6 Conclusions and Implications

In an economy where the divergence of metropolitan and urban markets is increasing, the market requirements for firms in economically disadvantaged areas are also greater. Here, for SMEs and the regions they belong to, knowledge and innovation are important tools for their economic development. As stated at the outset, the objectives of this study were to identify the influence of knowledge and innovation in SMEs in disadvantaged regions, and to show how these two phenomena can give SMEs sustainability, as well as developing their regions. In order to fulfil these objectives, six exploratory case studies were conducted in the Beira Interior Region of Portugal.

This study concluded that knowledge and innovation are seen to be relevant tools for the growth and sustainability of SMEs and consequently the regions where they are located. The importance for the regions is mentioned in all the cases studied here, since value is created in them through knowledge and innovative activities. Employment is created and purchasing power is increased. Innovation also leads to other services, reflecting a region's favourable image in relation to others. When companies set up in disadvantaged regions, the number of collaborators can increase, as does the population.

The results also show that focusing on knowledge and innovation can be a route to firms' survival and success, as this allows the exploitation of new markets, new products, equipment or processes, maintaining competitiveness at the national level and giving access to international markets. In the case of the firms studied, their strategies rely greatly on the capacity to manage knowledge, entrepreneurship and innovation as a way to improve the quality of their products and provide a wider supply. In these circumstances, SMEs extend their range of clients, expand into new markets, increase sales, differentiate themselves from the competition and improve their organisational performance.

Table 4.3 Comparative analysis of cases

Cases/ firms	Knowledge and innovation	Economic and regional development
1	*How*: New products, market expansion, product image and equipment image *Difficulties*: Financial and bureaucracy	*Performance indicator*: Product or service quality. Increase production, increase employees' production capacity *Development*: Bring more people to the regions, who will develop other projects and activities, with new ideas
2	*How*: Product differentiation, market expansion and production methods *Difficulties*: Bureaucracy	*Performance indicator*: Consulting clients. Increase firm wealth, create new markets, increase collaborators and clients *Development*: With increased sales, the number of collaborators will grow, increasing the region's population. This has an impact on the region as new business is developed, and other firms are also more attracted to the region
3	*How*: Decoration, product image, new products and market expansion *Difficulties*: Lack of a qualified workforce and suppliers in the region	*Performance indicator*: Looking to the international market and higher quality *Development*: a firm like this, which brings raw material from outside, produces here and sends away again, does not generate regional wealth
4	*How*: Internal organisation and technology *Difficulties*: Bureaucracy and justice	*Performance indicator*: Product or service quality; increase the firm's wealth, create new markets, increased collaborators and clients *Development*: It is firms that stimulate the region, through job creation
5	*How*: New products, new technology, production methods and partnerships *Difficulties*: Time to respond to clients	*Performance indicator*: Net result at the end of the year *Development*: Socially for the regions, since "value is created through entrepreneurship"
6	*How*: New computer system, new products, market expansion and infrastructure *Difficulties*: Access to finance and adapting to changes	*Performance indicator*: Fulfilment of objectives *Developments*: Create more employment, which generates commerce and the money in circulation. Also has a social value

Source: Own eleboration

It is also highlighted that this research can provide information and serve as benchmarking for companies or regions that choose knowledge and innovation strategies as methods leading to more efficient performance, as well as the development of their regions. For managers, business people and entrepreneurs whose firms are located in economically depressed regions, some suggestions/recommendations are now presented:

(a) Due to the difficulty in obtaining finance, firms could seek other sources of funding outside the banking system, looking for support from business associations that have the relevant knowledge.
(b) Supporting or developing areas of innovation and knowledge in firms can create new opportunities.

(c) Promoting partnerships with other firms, universities and associations, to support innovation and the share of knowledge, are ways to identify new opportunities, services, processes and knowledge.
(d) Increasing formal and informal training, facilitating collaborators' access to the subjects of knowledge management and innovation, especially in SMEs in disadvantaged regions, is a strategy to consider.
(e) Attracting possible collaborators with talent and new knowledge in the areas of innovation or marketing could solve or minimise some difficulties faced by the lack of a qualified workforce.

This study is not without limitations. Since the methodology adopted is related to qualitative analysis, one limitation is the fact that the results cannot be generalised. The study sample consisted of six SMEs, generating data only for these cases. A greater number of firms from different regions and sectors of activity could reveal different data, also increasing the reliability of the studies and obtaining more definite results.

Having confirmed the importance of knowledge and innovation not only for the firms but also for the regions they belong to, it will be pertinent to carry out other studies in these fields in the future. Since this research was based on a specific region of Portugal (Beira Interior), it would also be interesting to explore the impact of knowledge and innovation in other regions or countries, allowing a comparative analysis to be made, in order to determine the significance of those geographical and even cultural differences, and develop new strategies and policies to support the development of firms located in disadvantaged regions.

Acknowledgements The authors gratefully acknowledge financial support from FCT and FEDER/COMPETE through grant PEst-C/EGE/UI4007/2013.

References

Acs, Z. J., & Armington, C. (2004). Employment growth and entrepreneurial activity in cities. *Regional Studies, 38*(4), 911–927.

Acs, Z. J., & Armington, C. (2006). *Entrepreneurship, geography, and American economic growth*. Cambridge, MA: Cambridge University Press.

Acs, Z. J., & Storey, D. (2004). Introduction: Entrepreneurship and economic development. *Regional Studies, 38*(8), 871–877.

Acs, Z. J. (2008). Foundations of high impact entrepreneurship. *Foundations and Trends in Entrepreneurship, 4*(6), 535–620.

Asheim, B. T., Boschma, R., & Cooke, P. (2011). Constructing regional advantage: Platform policies based on related variety and differentiated knowledge bases. *Regional Studies, 45*(7), 893–904.

Audretsch, D. B., & Fritsch, M. (2002). Growth regimes over time and space. *Regional Studies, 36*(2), 113–124.

Audretsch, D. B., & Keilbach, M. (2004). Entrepreneurship capital and economic performance. *Regional Studies, 38*(8), 949–959.

Audretsch, D. B. (2003). *Entrepreneurship: A survey of the literature*. Enterprise Papers No. 14. Brussels: Office for Official Publications of the European Communities.

Bartlett, W., & Bukvić, V. (2001). Barriers to SME growth in Slovenia. *Economic Policy in Transitional Economies, 11*(2), 177–195.

Baumol, W. J., Litan, R. E., & Schramm, C. J. (2007). *Good capitalism, bad capitalism, and the economics of growth and prosperity*. New Haven: Yale University Press.

Bruton, G. D., Ahlstrom, D., & Obloj, K. (2008). Entrepreneurship in emerging economies: Where are we today and where should the research go in the future. *Entrepreneurship Theory and Practice, 32*(1), 1–14.

Carlino, G., & Kerr, W. R. (2015). Agglomeration and innovation. In G. Duranton, J. V. Henderson, & W. C. Strange (Eds.), *Handbook of regional and urban economics* (Vol. 5, pp. 349–404). Boston, MA: Elsevier.

Carree, M. A., & Thurik, A. R. (2003). The impact of entrepreneurship on economic growth. In Z. J. Acs & D. B. Audretsch (Eds.), *Handbook of entrepreneurship research: An interdisciplinary survey and introduction* (pp. 437–471). Dordrecht: Kluwer Academic Publishers.

Chandler, A. D. (1962). *Strategy and structure: Chapters in the history of the industrial enterprise*. Cambridge: MIT Press.

Dijkstra, L., Annoni, P., & Kozovska, K. (2011). *A new regional competitiveness index: Theory, methods and findings*. European Union Regional Policy Working Papers, No. 02. Brussels: European Commission.

Drucker, P. F. (1985). *Innovation and entrepreneurship: Practice and principles*. New York: Harper & Row.

Eisenhardt, K. M. (1989). Building theories from case study research. *Academy of Management Review, 14*(4), 532–550.

Elberse, A., & Eliashberg, J. (2003). Demand and supply dynamics for sequentially released products in international markets: The case of motion pictures. *Marketing Science, 22*(3), 329–354.

European Commission. (2003). Commission recommendation of 6 May 2003 concerning the definition of micro, small and medium-sized enterprises. *Official Journal of the European Union, 361*(124), 36–41.

Florida, R. (2003). Entrepreneurship, creativity, and regional economic growth. In D. M. Hart (Ed.), *The emergence of entrepreneurship policy: Governance, start-ups, and growth in the U.S. knowledge economy* (pp. 39–60). Cambridge, MA: Cambridge University Press.

Florida, R. (2005). *Cities and the creative class*. New York: Routledge.

Fritsch, M., & Müller, P. (2007). The persistence of regional new business formation-activity over time – Assessing the potential of policy promotion programs. *Journal of Evolutionary Economics, 17*(3), 299–315.

Gries, T., & Naudé, W. (2010). Entrepreneurship and structural economic transformation. *Small Business Economics, 34*(1), 13–29.

Gupta, P. D., Guha, S., & Krishnaswami, S. S. (2013). Firm growth and its determinants. *Journal of Innovation and Entrepreneurship, 2*(15), 1–15.

Kamalian, A. R., Yaghoubi, N. M., & Elyaskordi, A. (2011). Entrepreneurship development in the organization and its role in entrepreneurship strategy. *European Journal of Humanities and Social Sciences, 1*(1), 51–65.

Kim, P. H., Aldrich, H. E., & Keister, L. A. (2006). Access (not) denied: The impact of financial, human, and cultural capital on entrepreneurial entry in the United States. *Small Business Economics, 27*(1), 5–22.

Kofjač, D., Knaflič, A., & Kljajić, M. (2010). Development of a web application for dynamic production scheduling in small and medium enterprises. *The Organ, 43*(3), 125–135.

Maillat, D. (1992). Milieux et dynamique territoriale de l'innovation. *Canadian Journal of Regional Science, 15*(2), 199–218.

McMillan, J., & Woodruff, C. (2002). The central role of entrepreneurs in transition economies. *Journal of Economic Perspectives, 16*(3), 153–170.

McMillan, M. S., & Rodrik, D. (2011). Globalization, structural change and productivity growth. In M. Bachetta & M. Jansen (Eds.), *Making globalization socially sustainable* (pp. 49–84). Geneva: International Labor Organization.

Moore, S. B., & Manring, S. L. (2009). Strategy development in small and medium sized enterprises for sustainability and increased value creation. *Journal of Cleaner Production, 17*(2), 276–282.

Naudé, W. (2013). *Entrepreneurship and economic development: Theory, evidence and policy.* IZA Discussion Paper No. 7507.

O'Gorman, C. (2001). The sustainability of growth in small and medium-sized enterprises. *International Journal of Entrepreneurial Behavior & Research, 7*(2), 60–75.

OECD. (2008). *Promoting entrepreneurship and innovative SMEs in a global economy.* Paris: OECD Publishing.

Omar, S. S., Arokiasamy, L., & Ismail, M. (2009). The background and challenges faced by the small medium enterprises. A human resource development perspective. *International Journal of Business and Management, 4*(10), 95–102.

Pillania, R. K. (2006). Leveraging knowledge for sustainable competitiveness in SMEs. *International Journal of Globalisation and Small Business, 1*(4), 393–406.

Popova, V., & Sharpanskykh, A. (2010). Modeling organizational performance indicators. *Information Systems, 35*(4), 505–527.

Porter, M. E. (1990). *The competitive advantage of nations*. New York: Free Press.

Porter, M. E. (2000). Location, competition, and economic development: Local clusters in a global economy. *Economic Development Quarterly, 14*(1), 15–34.

Praag, C. M. & Versloot, P. H. (2007). *What is the value of entrepreneurship? A review of recent research.* IZA Discussion Paper No. 3014.

Saxena, S. (2012). Problems faced by rural entrepreneurs and remedies to solve. *Journal of Business and Management, 3*(1), 23–29.

Schwab, K. (2014). *The global competitiveness report 2014–2015.* Geneva: World Economic Forum.

Stam, E., & Wennberg, K. (2009). The roles of R&D in new firm growth. *Small Business Economics, 33*(1), 77–89.

Szirmai, A., Naudé, W., & Goedhuys, M. (2011). Enterpreneurship, innovation, and economic development: An overview. In A. Szirmai, W. Naudé, & M. Goedhuys (Eds.), *Enterpreneurship, innovation, and economic development* (pp. 3–32). Oxford: Oxford University Press.

van Stel, A., Carree, M., & Thurik, R. (2005). The effect of entrepreneurial activity on national economic growth. *Small Business Economics, 24*(3), 311–321.

Wennekers, S., van Wennekers, A., Thurik, R., & Reynolds, P. (2005). Nascent entrepreneurship and the level of economic development. *Small Business Economics, 24*(3), 293–309.

Yin, R. K. (2013). *Case study research: Design and methods* (5th ed.). Los Angeles: Sage.

Chapter 5
Applied Innovation Methodology: A Proposal for a Dynamic Sustainable Environment for the Generation of Innovation and Knowledge Management Practices in SMEs

Jordi Mauri-Castello, Antonio Alonso-Gonzalez, and Marta Peris-Ortiz

5.1 Introduction

Over time people have evolved to create companies in order to produce goods and services to satisfy their needs. Historical development has shown that companies have adapted to changes in their environment and therefore to the changing needs of the people because these diversifications are generated by their expectation. There are various concepts of evolution, but it is considered interesting to take it into account as an internal mechanism to improve the transmission to living descendants. Given the complexity of the issues involved, multiple methods, procedures, and the mechanisms used, they are configured for each case according to the particular conditions (Bank Boston 1997). This line of thought is fully aligned with the Schumpeterian system which is based on a necessary symbiosis between economic, historical, political, social, and other elements related to the development of the capitalist system (Schumpeter 1934), as well other approaches that are conditioned by the analysis of the innovation concept, for example, the organizational knowledge creation (Nonaka 1994), the resource-based perspective (Peteraf 1993; Barney 2001), or the sustainability-driven approach (Teece 2007; Kiron et al. 2013).

J. Mauri-Castello
Institute of Plastics Engineering, Universität Stuttgart, Stuttgart, Germany
e-mail: jordi.mauri@ikt.uni-stuttgart.de

A. Alonso-Gonzalez (✉)
EIAM—PRIME Business School, Universidad Sergio Arboleda, Bogotá, Colombia
e-mail: antonio.alonso@usa.edu.co

M. Peris-Ortiz
Department of Business Administration, Universitat Politècnica de Valencia, Valencia, Spain
e-mail: mperis@doe.upv.es

M. Peris-Ortiz et al. (eds.), *Knowledge, Innovation and Sustainable Development in Organizations*, Innovation, Technology, and Knowledge Management,
https://doi.org/10.1007/978-3-319-74881-8_5

In this paper many conceptual works found in the literature dealing with the evolution of innovation were described (Tidd et al. 2005), as well as knowledge (Prahalad and Hamel 1995), ICT (information and communications technologies) (Prahalad and Hamel 1995), organizational capabilities, new organizational structures, human capital, absorptive capacity, and information society (Moore 1999). It has also been explained how these factors affect the adaptation of companies in a globally competitive environment (Fernández de Lucio et al. 1996), rapid technological change, and innovative approaches, achieving significant changes in the competitive and productive capacity of the company (Porter 2008).

After performing a literature review, this study searches to bridge the existing gap between corporate organizational structure and innovation (Barceló 1994) and other new concepts relating to creativity management (Davenport 1993). Therefore, a new model to allow companies to implement more effective, efficient, and competitive was proposed. Some authors like Escorsa and Valls (1997) argue that "the scope, flexibility and efficiency of the knowledge assimilation depends on the company's organizational structure (functional, divisional or matrix)." Other authors explain that we must consider companies that have to hire subcontractors, freelancers, or consultants in order to acquire specific skills, creating a network of subcontractors, leading a generation of strategic alliances of individuals (that are not personally interconnected, but located globally). Perhaps these actors do not know each other personally, due to the fact that they could be located at various locations in the world. Co-working has the same need for this perspective and can also be considered as a network of subcontractors.

Thus, this work aims to deepen and strengthen the scope of knowledge management and its relationship with the innovative activities of the company. Specifically, the main feature of managing organizational knowledge has been analyzed, and the relationship between human behavior and innovative thinking, identified in relation to the specific correlation of business results. Moreover, other studies have been taken into account (Freeman 1987), which shows the importance of thinking and linking other perspectives to the organizational approach and the importance of increasing the balance in three different areas: the commercial and economic dimension, the demographic movements, and the new technological scenarios.

5.2 Theoretical Background

5.2.1 Conceptual Framework: Flexibility, Organizational Design, and Knowledge Management

According to Moore (1999), "companies must have the ability to adapt to new situations, without these changes they entail heavy penalties of time, cost, effort or performance, a concept which is known as flexibility. Flexibility is associated with the human capital that owns the company and the orientation adopted by certain practices of human resource management in terms of job design and selection, training and development, performance evaluation and compensation." Thus, flexibility is an important concept to take into account, if it is desired to implement an innovative environment within a company. However, some authors like Christensen and Raynor (2003) assume that organizational context influences the level and frequency of innovative behavior of the company, defining the innovative capacity concept as the ability to successfully implement new ideas. They measure this capacity in accordance with the mechanisms that favor the flow of innovative ideas, including organizational structure and management ideas.

The concepts of flexibility, organizational context, or structure were described, as well as its importance and relationship with the innovative capacity or innovation processes of a company. The glue that must link together these terms is the company's knowledge. Following Nonaka and Takeuchi's (1995) work, knowledge has been described as the information that has been subjectively contextualized, interpreted, and assimilated by an individual, group, or organization. Knowledge includes information but also competencies, skills, and abilities to process and interpret this information. The fundamental priority is to understand the nature of knowledge and how it is generated. Knowledge management is necessary to deal with the turbulence of the current business environment, giving the company a level of flexibility necessary for adaptation. Consequently, knowledge must be created, developed, transferred, and applied to products and/or services, and this process can be favored by the organizational culture, design, and human resources management practices.

According to Nonaka and Takeuchi (1995), "the distribution of knowledge in the organization is a central criterion in the organizational design and this knowledge resides primarily in people, and given the bounded rationality it is not possible to concentrate all the relevant knowledge for decision making in a single person. Information and knowledge are relevant in the decision process, so that when making the assignment of decision responsibilities between the members of an organization it is necessary to consider the nature of the relevant information, where and how it is produced, and the cost of transferring efficiently." Knowledge management allows to analyze, to organize, and to implement the information in a business context in order to convert it into an intangible added value for the organization, allowing decision-making and corporate management to become easier.

5.2.2 *Innovation, Change Management, and Corporate Culture*

In words of Fernández de Lucio (1991), "innovation is often closely connected with a company founder's vision and entrepreneurial spirit. Sometimes it is also driven by people who have a predisposition to change, the so-called masters of change. These individuals are not always great experts, and they are not necessarily more creative than others. The truth is that they are more inclined to break free from traditional knowledge and deal with issues from a different angle, being considered more persistent in order to defend new projects more passionately. They know how to communicate their ideas and get key players within the organization to assist them in the process of formulating them. However, people alone cannot create a culture of innovation." Other authors are focused on how open innovation is defined and the ways in which the construct is used in a variety of empirical research settings (Dahlander and Gann 2010).

In regard to corporate culture, successful companies know that it is an important competitive factor, though this concept may be rather difficult to assimilate by certain individuals within the company. It is important to highlight that it is relatively easy to imitate products, services, and processes, but corporate culture is unique and it persists over time, being a corporate tool that not every organization can create. Scott (1989) mentioned that "those companies whose employees are open to new ideas and are prepared to actively take part in technological change by being creative and showing initiative usually have greater potential for innovation. However, a corporate culture does not consist of people only. It is the systems, procedures and organizational structure which lead the individuals to interact in a certain way. Those companies managing highly complex projects, in which the high content of new features requires them to take advantage of all the specialist knowledge available, have had to turn to new organizational models. One of the main differences between traditional and strategic, process-based management is how responsibility is assumed and understood." Scott (1989) continues arguing that "this often leads companies to abandon their horizontal structure with a fixed hierarchy, in favor of a matrix based structures with a positional hierarchy in which the same person can be in charge of a process that extends to a number of different areas. For this reason, the innovation process manager is a key player that needs the total trust and commitment of management. In small businesses this position is held by the top director."

5.2.3 Innovation and SMEs

Innovation has become a key process to achieve competitive advantages, and it must be conducted when efforts to improve it have reached their limit and are no longer sufficient to increase efficiency, so new ways are sought to develop processes, products, or services in order to make fundamental or substantive changes and advances. Innovation must be understood as a core resource which does not guarantee that competitiveness will be reached, but it is necessary to establish methodologies and strategies that lead to achieve this objective. The studies on the factors involved in the process of innovation, as well as opportunities in the different scenarios, provide insights into potential tools involving innovation as part of the system within the organization and therefore open up new options for competitiveness (Armenta et al. 2016).

If a focus on SMEs is considered, it must be said that several attempts have been made to build models of innovation for companies, but there are no documented empirical studies showing how SMEs have achieved and capitalized their experience and knowledge in terms of innovation. There can be found some studies relating to factors that determine innovation such as organizational creativity, entrepreneurship, intellectual capital, knowledge management, organizational management, business incubation, and continuous improvement, but they are isolated works that do not allow a model of innovation for small and medium enterprises to encourage their permanence and growth in the market (Parra et al. 2016). Thus, in reference to SMEs, innovation can be described as a complex technological, sociological, and economic process that involves large interactions, both within the company and between the company and its economic, technical, competitive, and social environment. Therefore, it cannot be expected to be a success or to be explained satisfactorily in terms of just one or two factors, but as a group of various factors, closely interrelated, which must work together to create and reinforce the kind of environment that facilitates the success of technological innovation. No single element is likely to be effective and, therefore, no management tool or technique will create and sustain an enabling environment for innovation (Armenta et al. 2016). Nowadays, an increasing number of SMEs are incorporating new information technologies and communications systems within their processes, so it can be expected that the level of innovation and creativity in these SMEs will be increased in the coming years; therefore, their level of performance will encounter positive changes, thereby obtaining additional resources to address financial problems (Guzman 2016).

In another way, the collaborative activities undertaken by organizations, mainly SMEs, with companies and public and private institutions not only generate greater innovation activities but also create the necessary conditions and best practices for the adoption and implementation of innovation. Cooperation should be considered by managers of SMEs not only as a business strategy, but rather as an activity of everyday life in companies, which will be implemented in all departments or functional areas of the organization, so that it will be feasible to achieve the benefits generated by these collaborative activities. If organizations, especially SMEs, want to adopt or increase their innovation activities, then it will be necessary for managers of SMEs to seek the implementation of innovation activities, with other companies and public and private organizations. Through collaboration with other organizations, there is a distinct probability that SMEs will obtain a higher level of innovation. Therefore, cooperation will be implemented in companies before innovation activities, because in this way the human, technical, and financial resources available to organizations will have improved results and an impact in the innovation level of SMEs (Lopez-Torres et al. 2016). SMEs should create a work environment within the organization that promotes and encourages the collaboration of workers and employees, so that the staff is motivated to participate in collaboration activities through teamwork, which could be supported by external staff. In addition, companies also have to implement this collaborative system with their customers and suppliers in the decision-making of the organization. Otherwise, SMEs can find it difficult to defend a position of power in relation to customer satisfaction, production levels, and marketing tasks in reference to their products and services commitments. Thus, collaborative activities with its customers and suppliers will be essential to significantly improve their level of innovation (Lopez-Torres et al. 2016). The creation of collaborative networks between corporate universities, customers, and SMEs could be another mechanism of value creation according to Alonso-Gonzalez et al. (2017b).

Innovation is a competitive factor whose strategic value has been questioned by some experts through the literature review. However, getting a profitable exploitation of innovation is a difficult and complicated task. This complexity turns out to be even greater when the case of start-up companies is analyzed. Innovation involves risk out of a comfort zone, and this decision involves a lot of uncertainty, for individuals and organizations that wish to undertake this task. Therefore, it is important that from an early stage these start-ups develop a capacity to absorb knowledge as a distinct ability to explore new opportunities (products, services, companies) and entrepreneurial orientation focused on the market, for the successful exploitation of such knowledge (production and marketing of new products and services). Conditions, absorption capacity and entrepreneurial orientation, are essential to receive profit from the effort that targets innovation (Díaz et al. 2015).

5.2.4 Innovation and Entrepreneurship in Profit and Nonprofit Organizations

In today's economy focused on a world in the process of postmodernity, mankind has the need to look for opportunities that generate added value and economic benefits and accelerate the life cycles of other enterprises, as a survival factor of their organizations in our current competitive climate. This phenomenon of entrepreneurship and creation of organizations has been an engine of development in social and economic processes, which have brought humanity to an evolution. One of these developments has been the technological invention, economic growth, and new forms of production as a result of this business development. This whole process of change has had different stages, from the modes of production, feudalism, capitalism, and the industrial revolution toward a knowledge economy and new products, services, and process developments brought to our current modernity (Ballestas 2016).

It is remarkable to highlight the correlations between entrepreneurship and economic development. For example, in the study performed by De Oliveira and Cardoso (2015), it can be observed specifically the relationship between economic and social development and the level of entrepreneurial activity in the world. The authors established two research hypotheses and estimated four econometric models. The first hypothesis expected a negative relationship between entrepreneurship and economic, social, and competitiveness development of a country. The hypothesis was confirmed, meeting previous studies and showing that countries with high rates of entrepreneurship do not necessarily obtain social economic development. These interesting results suggest that there is an opportunity for countries to establish conditions and structures for the enhancement of entrepreneurship and consequently promote higher levels of economic and social growth, as well as an increase in competitiveness.

It is important for entrepreneurs to seek and provide interesting and modern alternatives related to innovation in order to generate new ways to compete and be profitable in our current highly demanding market; develop new products, dynamic services, and optimized processes; and provide a competitive advantage to allow them to consolidate in the market (Roman 2016). There is a relationship between social entrepreneurship and social innovation. From the point of view of the entrepreneur, it must be highlighted that the role of innovation is a way to achieve social goals. Thus, according to this idea the entrepreneur should aim to create profit but also to generate social value. It is necessary to show how the use of innovative practices is able to break with existing paradigms, and these practices can make businesses as a benchmark for society and for other companies. The situation would not be uncommon to public institutions, which should modify their regulations and adapt to the new economic and social reality around them. This way, a social change would not only be achieved but also a new conception of the entrepreneur concept (Alonso et al. 2015). In this line of thought, some studies can be found, for example, the one performed by Molina-Betancur and Polanco-López de Mesa (2015), exploring the possibilities of how public institutions could encourage the development of science, technology, innovation, and entrepreneurship through the theoretical relationship between government and sustainability. Another study which follows this line of thought is one performed by Alonso-Gonzalez et al. (2017a), describing collaborative initiatives within universities and a population at risk of social exclusion.

Molina and Velilla (2016) found in another study strong correlation for the importance of innovation in entrepreneurship, describing that business analysis should consider such variables relating to innovation, which have often been underused. Moreover, the results of their work suggest a business channel driven by innovation and furthermore due to synergy among individuals, similar to the intergenerational effects. From the point of view of economic policy, entrepreneurship is currently a work activity that is being encouraged by economic, social, and political institutions. It is mandatory to identify the most important factors that engage individuals, facilitating the work of guiding these incentive policies and improving their effectiveness and efficiency.

The role of higher education institutions is also an important consideration, describing them as catalysts of entrepreneurship, innovation, and growth for economic development, especially in developing countries, in reference to the studies conducted by Alonso-Gonzalez et al. (2016) and Alonso-Gonzalez et al. (2017a, b, c). These institutions could be seen as innovation generators and large technology centers that could enhance innovation, creativity, and entrepreneurship within society, as well as technological development. Some of the products and services associated with the infrastructures are project incubators, qualified training in various areas for students in business entrepreneurship, research of academic projects focused on the student community and the general public, technological development, promotion of technological transfer to the business market, management and generation of business knowledge in training students who graduate through the creation of these specialized centers, and start-ups or business creation initiatives (Sinisterra 2016).

For this social entrepreneurship through higher education institutions, emphasis must be placed on the promotion of knowledge and Internet applications, as well as directives, accounting and technical competitiveness, managerial autonomy and risk management, acceptance of error and decision-making in SMEs, copyright and intellectual property concepts, commercial exploitation, and transfer technology. For infrastructure in relation to innovation and entrepreneurship, the importance of university institutions must be highlighted (quantity and quality of these centers), as well as the acquisition of research tools, techniques and business consolidation incubators, and technology-based and service centers for entrepreneurs. On the issue of entrepreneurial characteristics for innovation, the following attributes should be included: initiative, leadership, problem solving, and acceptance to change. For entrepreneurship discipline, these include the following: independence, risk, problem solving, communication, courage, self-control, and creativity. For both innovation and entrepreneurship, the attention should be focused on the following features to promote them transversely: the learning of creativity, problem solving, management risk, the use of ICT, the development of knowledge networks, self-learning and lifelong learning with a research and scientific approach, as well as the creation of institutional or corporate bonds (Herrera and Álvarez 2015). It is important to outline the burden of learning in this innovation and entrepreneurship dimension, which is perfectly aligned with the work of Crossan et al. (1999) concerning the organizational learning framework.

5.3 Methodology

5.3.1 Innovation and Entrepreneurship in Profit and Nonprofit Organizations

Once the review of the present literature has been highlighted in the current paper, and derived from the model proposed by Terre-i-Ohme (2002) from his document *Guide for Managing Innovation: Part 1, Diagnosis*, a model called Applied Innovation Methodology will be proposed as a tool to develop innovation and entrepreneurship in profit and nonprofit organizations and especially in SMEs. This methodology consists of six different phases which are conceptualization, development, applied innovation, optimization, development, and sustainability. These stages will be described in the following sections.

5.3.2 *Conceptualization: Setting a New Proposal of the Applied Innovation Methodology*

Through the Applied Innovation Methodology, the required human resources were established, as well as the technical processes and the methodological environment necessary to encourage the fulfillment of new concepts in order to guarantee success in the process. Therefore, the following issues are the beneficial results of this methodology:

- Creation of distance between the business and its competitors
- Anticipation to future competition moves and counterstrikes
- Creation or consolidation of new market places
- Increase in the existing area of business potential
- Generation of new business models and ideas

5.3.3 *Development: Qualitative Definition*

It is proposed that the Applied Innovation Methodology is the leader in the innovation process of an SME and consequently stabilizing the specifications, requirements, and processes in order to transform an idea into a tangible result. At this stage it is important to introduce the task that is performed by the so-called Innovation Team. This team in any SMEs consists of a group of experienced professionals in various fields such as design, engineering, logistics, production, finance, and marketing. The tasks of the so-called Innovation Team in the area of development need to be focused on the definition of the qualitative aspects of the product, specifying processes to optimize resources while looking to capitalize opportunities offered by the target market. The most important tasks that the Innovation Team must manage are the following:

- Implementation: in this process, technologies, industrial processes, and commercial requirements of costs, product launch, and presentation are defined. All these factors reflect strategic and technical values defined in the previous phase, allowing for the evolution, validation, and solidification of tangible elements throughout the process of a product or service.
- Strategic planning: the actual complexity of the production processes and the importance of factors like globalization and delocalization of companies and industrial processes require a correct planning of the "What, How, Where and How." A correct definition of the necessary processes to physically implement and launch a new product must establish the required frame to define new business opportunities.
- Commercial tools: the innovation process needs a team that actively helps the client to expose all product values to the user and collaborates with the sales force to maximize the development of tools that correctly transmit the approaches

developed in new products. The Applied Innovation Methodology has developed different commercial tools to specify all the factors that allow us to extract the value of all concepts.

5.3.4 *Applied Innovation: Performing Change and Growth Under Control*

Innovation should not be implemented in isolation, because of its high correlation within the specific needs of markets and the environment and other strong relationships related to the company's sector, products, and knowledge, as well as its limitations. The Applied Innovation Methodology aligns the client's knowledge with the strategic vision of the company, creating a new approach toward the knowledge management of the organization. Based on this existing knowledge, the Applied Innovation Methodology allows the Innovation Team project to define a new competitive scenario within the company for the client and the market.

Any company has a different and variable knowledge life cycle with its own expiration date. Besides, as the market changes, evolves, and modifies its needs and requirements, businesses often cannot develop the capacity and flexibility to adapt. The Applied Innovation Methodology overcomes this trend giving it another approach and initiating a process which identifies new opportunities and competitive positions within a market, generating new business opportunities. This new and innovative approach is not always produced at the same stage, nor does it produces changes in the same way in every situation Therefore, the innovation levels can be established depending on the needs of an organization:

- Strategic level: business strategy modification or adaptation
- Product range level: new business units or new products and services ranges
- Products conceptualization: new products and services functions or solutions
- Product attributes: new product presentations and new products and services external perceptions

5.3.5 *Optimization: Improving the Profitability of the Product*

Before the final version of the product is defined, it is necessary to perform a study of the elements which influences in the development process in order to achieve two essential objectives:

- Reduce final costs: contributing to innovative and constructive solutions and new materials, rationalizing the product range and always bearing in mind the technological availability and output capacities, or the optimization of commercial references.

- Improve competitiveness: the relationship between commercial diversity and technological platforms will determine the concept, form, investments, and final product price. A correct relation between both, which is defined in the first state of the project, will be vital to achieve its aims.

The optimization initiatives developed using the Applied Innovation Methodology create new feasible commercial initiatives according to the identified needs of the market, maximizing on its way different economic factors related to the product and making an impact in different areas in the final configuration of the product or service: design, logistics, production, quality control, marketing, and sales. These lines of optimization that should be taken into account are the following ones:

- Definition of platforms and modules: creating little infrastructure which then permits the maximization of the number of references that adapt to the needs of the client, minimizing the final cost of the product
- Alternative technologies: identifying materials and technology that reduce the final cost while maintaining or improving attributes of the product
- Product ranges: proposing an optimal point between commercial range (product references), needs of the market, and development in the function of commercial references

It is important to highlight that there are two key factors to take into account in this optimization process: to always maintain a relationship between value and cost from the market needs' point of view and to propose the best technical solutions from the product value's point of view, taking into account the materials and production processes, the suppliers efficiency and availability, as well as the Innovation Team expertise.

5.3.6 Development: Materializing Ideas in Products and Solutions

The Applied Innovation Methodology combines both engineering and design dimensions with a strategic vision, enhancing decision-making at the various strategic areas within an organization. This strategic vision allows the Innovation Team of the SME to define unique competitive elements that are extremely difficult to copy or implement by the competition, allowing the creation of unique competitive concepts and proposals, designing and creating resources which will add knowledge and added value to the different areas of the company, as well as to the final customer or target market. The combination and alignment of these two different ways of understanding the product result in a knowledge increase without a loss of value proposal. However, it is difficult to strike this balance and for the competition to replicate this new value proposal in the target market and this is the main strength that the Applied Innovation Methodology achieves, transforming these ideas into business opportunities for the customers in the following ways:

- Highlighting the attributes oriented to the final user's needs with an attractive design
- Optimizing for any product its functional values, material specifications or requirements, technological functionality, or mechanics efficiency
- Adapting or modifying the manufacturing process to meet specific customers' needs relating to design, ergonomics, or product performance

In every stage within the development process, there is a validation exam set up by a multidisciplinary team which evaluates if the product is achieving the specifications required by the client. Therefore, the Applied Innovation Methodology establishes the link between the strategic objectives, differential values, and value proposal and commercial initiatives before the launch of the product, with cost, quality, and time frame defined by the market.

5.3.7 Sustainability: Taking into Account Responsibility and Ethics

Sustainability can be defined as a mix of social benefits for mankind, economy, and environment in order to improve the present and future quality of life. The Applied Innovation Methodology takes into account this concept in its development and production process of its products and services, aligning this with its objective of developing competitive advantages in the markets. These three different dimensions of the sustainability concept are interconnected and the Applied Innovation Methodology's points to minimize every harmful effect and encourage positive outputs on each of the following:

- Environment: the environment has a direct effect on quality of life and social wellness, and the natural resources it offers provide the substance from which business is created and a sustainable economy works. The implementation and enforcement of international standards and legislation now recognize the need for business actions and product improvements in an environmentally friendly way (EU producer responsibility legislation, IPP (Integrated Product Policy), ISO 14000, eco-labeling, etc.).
- Society: businesses are beginning to be held accountable for their actions in terms of the welfare of employees and citizens. International regulations, audits, reports, and evaluations are becoming commonplace, which have implications in products and services (Corporate Social Responsibility, Dow Jones Sustainability Index, Global Compact, Global Reporting Initiative, etc.).
- Economy: as a bottom line, it is essential that responsible and environmentally improved products remain competitive. In the marketplace, a win-win situation occurs when environmental and economic savings are achieved through one ini-

tiative, as, for example, the search of efficiency and optimization. Furthermore, product responsibility has direct positive effects on financial results, and in some cases certification is a requirement in the market (Forest Stewardship Council certification in furniture, ISO 14000 standard for suppliers, etc.).

The focus of the Applied Innovation Methodology, considering these different dimensions of the sustainability concept, is to increase the added value all along the entire life cycle of products and services, describing this process through the following stages:

1. Specification of materials: minimization of quantity of materials and rational application of recycled, recyclable, biodegradable, and renewable materials
2. Efficiency in manufacture: specification of the manufacturing process to generate the least amount of possible waste and optimization of the number of processes and energy used in each stage
3. Optimization in distribution: improvement of packaging and rationalization of the logistics and transport to avoid inefficiencies.
4. Use: evaluation and reduction of consumables and energy use throughout the worthwhile life of the product
5. End of life: consideration of waste materials generated throughout the products' life and definition of the process of recycling or reuse of products at end of their useful life

5.4 Conclusions and Future Research

As Drucker (1998) asks rhetorically "How can managers plan, let alone trust a process which, in essence, depends so much on creativity, inspiration and luck? […] Although some innovations are the result of a flash of genius, most of them, especially the most successful ones, are born out of a conscious and deliberate search for innovation opportunities that can only be found on rare occasions."

Innovation requires a merged vision and a multidisciplinary knowledge, and nowadays innovation processes have become a competitive advantage in an increasingly global economy, which cannot compete on the basis of reduced margins. A joint vision for all involved activities enables organizations to conceptualize and position their products or services in the markets. This is performed in a proper way beyond the opportunity costs of engineering. In our current competitive environment, new elements should be introduced to increase the balance in three different areas: the commercial and economic dimension, the demographic movements, and the new technological scenarios. Innovation is not an isolated issue, depending on the analysis, understanding, and satisfaction of needs for every client according to the environment, sector, products, and knowledge management. In this order of alignments, the process of conceptualization and the conception of an idea into a concept or prototype and then into a product or service with the required specifications are defined by the whole development process.

In the proposed Applied Innovation Methodology, an alternative source of detection and generation of new business opportunities has been described, facilitating the competitiveness of organizations based on the creation of new competitive spaces, using optimization to create a commercial range of rational product costs, consistent with consumer needs, maximizing the profitability. It has been identified in the model the importance of the selection of a key partner in order to reach the quality levels required and optimizing the product cost. Therefore, the Innovation Team should employ time and effort in this selection process of key partners, based on specifications and requirements of the project in order to fulfill the goals, and having considerable impact on the final product configuration in different areas such as design, logistics, production, quality control, marketing, and purchasing.

The Applied Innovation Methodology defined in the current work is a framework which helps to define the various methodologies, strategies, processes, and design costs and produce and market a product within its specifications and requirements. To perform this model, some external and private clients were consulted throughout the entire process in order to establish the best optimized route and therefore bring any product from the identified customers' needs to the target market with proper quality standards within a specified time frame.

References

Alonso, D., González, N., & Nieto, M. (2015). Emprendimiento social vs innovación social. *Cuadernos Aragoneses de Economía, 24*(1), 119–140.

Alonso-Gonzalez, A., Diaz-Morales, A., & Peris-Ortiz, M. (2016). Enhancement of entrepreneurship in Colombian universities: Competence approach plus personalized advice (CAPPA) model. In M. Peris-Ortiz, J. A. Gómez, F. Vélez-Torres, & C. Rueda-Armengot (Eds.), *Education tools for entrepreneurship* (pp. 101–112). Springer International Publishing, Switzerland. ISBN: 978-3-319-24655-0.

Alonso-Gonzalez, A., Palacios-Chacon, L. A., Rueda-Armengot, C., & Peris-Ortiz, M. (2017a). Collaborative networks between Colombian universities and population at risk of social exclusion: The Sergio Arboleda University experience. In M. Peris-Ortiz, F. Teulon, & D. Bonet-Fernandez (Eds.), *Social entrepreneurship in non-profit and profit activities: Theoretical and empirical landscape* (pp. 65–72). Springer International Publishing, Switzerland. ISBN: 978-3-319-50850-4.

Alonso-Gonzalez, A., Peris-Ortiz, M., & Mauri-Castello, J. (2017b). Collaborative networks between corporate universities, customers, and SMEs: Integrating strategy towards value creation. In M. Peris-Ortiz & J. J. Ferreira (Eds.), *Cooperative and networking strategies in small business* (pp. 197–205). Springer International Publishing, Switzerland. ISBN: 978-3-319-44509-0.

Alonso-Gonzalez, A., Plata-Rugeles, D., Peris-Ortiz, M., & Rueda-Armengot, C. (2017c). Entrepreneurial initiatives in Colombian universities: The innovation, entrepreneurship and Business Center of Sergio Arboleda University. In M. Peris-Ortiz, J. A. Gómez, J. M. Merigó-Lindahl, & C. Rueda-Armengot (Eds.), *Entrepreneurial universities* (pp. 151–163). Springer International Publishing, Switzerland. ISBN: 978-3-319-47949-1.

Armenta, J. R. C., Arango, O. M., & Gress, E. S. H. (2016). Medición de la innovación en las pymes mediante la aplicación de métodos multicriterio (electre y AHP). *Mercados y Negocios, 14*(2), 169–185.

Ballestas, C. L. (2016). Aspectos de innovación schumpeteriano: el emprendimiento, el perfil del empresario en el contexto social. *Dictamen Libre, 1*(17), 73–80.
Bank Boston. (1997). *MIT: The impact of innovation, Bank Boston Economics Department Special Report*. Boston, Bank Boston.
Barceló, M. (1994). Innovación tecnológica en la industria: Una persectiva española. *Llibres dels Quaderns de Tecnología*, Beta E. Editorial.
Barney, J. B. (2001). Is the resource-based "view" a useful perspective for strategic management research? Yes. *Academy of Management Review, 26*(1), 41–56.
Christensen, C. M., & Raynor, M. E. (2003). *The innovator's solution*. Boston: Harvard Business School Press.
Crossan, M. M., Lane, H. W., & White, R. E. (1999). An organizational learning framework: From intuition to institution. *Academy of Management Review, 24*(3), 522–537.
Dahlander, L., & Gann, D. M. (2010). How open is innovation? *Research Policy, 39*(6), 699–709.
Davenport, T. H. (1993). *Process innovation*. Boston: Harvard Business School Press 10-11, 50-55.
De Oliveira, I. G. S., & Cardoso, S. P. (2015). Emprendimiento y sus implicaciones para la innovación y el desarrollo: un análisis multivariante con los indicadores socioeconómicos. *Globalización, Competitividad y Gobernabilidad de Georgetown/Universia, 9*(2):43–60.
Díaz, Y., Guerrero, M., & Peña, I. (2015). Innovation driven productivity through corporate entrepreneurship/Productividad de la innovación a través del emprendimiento corporativo. *Universia Business Review, 15*(3), 33–47.
Drucker, P. F. (1998). The discipline of innovation. *Harvard Business Review, 76*(6), 149–157.
Escorsa, P., & Valls, J. (1997). Manual de Gestión e Innovación Tecnológica en la Empresa, *Ed. Centro Interuniversitario de Desarrollo CINDA, Colección Ciencia y Tecnología n° 41*, Santiago de Chile, 22–24, 29–30.
Fernández de Lucio, I. (1991). Industria: innovación y tecnología. In J. Honrubia, J. M. Bernabé, A. Ortí, & J. A. Tomás (Eds.), *La Comunidad Valenciana en l'Europa Unida Innovació i Tecnología* (Vol. VIII). Valencia: Generalitat Valenciana.
Fernández de Lucio, I., Conesa, F., Garea, M., Castro, E., Gutiérrez, A., & Bodegas, M. A. (1996). *Estructuras de interfaz en el sistema español de innovación. Su papel en la difusión de tecnología*. Centro de Transferencia de Tecnología. Universidad Politécnica de Valencia.
Freeman, C. (1987). *Technology and economic performance: Lessons from Japan* (pp. 4–25). London: Pinter.
Guzman, G. M. (2016). La influencia de las TIC y la innovación en el rendimiento de las pymes de Aguascalientes: una evidencia empírica. *Mercados y Negocios, 12*(1), 15–33.
Herrera, M. M., & Álvarez, J. G. C. (2015). Características para promover el emprendimiento y la innovación. Marco general para la formación en las universidades. UPIICSA. *Investigación Interdisciplinaria, 1*(2), 32–45.
Kiron, D., Kruschwitz, N., Reeves, M., & Goh, E. (2013). The benefits of sustainability-driven innovation. *MIT Sloan Management Review, 54*(2), 69.
Lopez-Torres, G. C., Guzmán, G. M., Castro, S. Y. P., & Ramírez, R. G. (2016). Colaboración y actividades de innovación en Pymes. *Contaduría y Administración, 61*(3), 568–581.
Molina, J. A., & Velilla, J. (2016). *La innovación como determinante pare el emprendimiento* [Innovation as determinant of entrepreneurship] (No. 71471). 1–27. University Library of Munich, Germany.
Molina-Betancur, C. M., & Polanco-López de Mesa, J. A. (2015). Un desafío de sostenibilidad: el nuevo distrito metropolitano de ciencia, tecnología, innovación y emprendimiento del valle del Aburrá en Colombia. *Estudios Geográficos, 76*(279), 579–607.
Moore, G. (1999). *Inside the tornado: Marketing strategies from Silicon Valley's cutting edge* (Revised ed.). New York: Harper Collins Publishers.
Nonaka, I. (1994). A dynamic theory of organizational knowledge creation. *Organization Science, 5*(1), 14–37.
Nonaka, I., & Takeuchi, H. (1995). *The knowledge-creating company: How Japanese companies create the dynamics of innovation*. New York: Oxford University Press.

Parra, M. E. L., Navarro, N. E. G., & Torres, M. D. C. V. (2016). Modelo de Innovación para las PYMES a partir de la práctica guiada. *Global de Negocios, 4*(7), 27–39.
Peteraf, M. A. (1993). The cornerstones of competitive advantage: A resource-based view. *Strategic Management Journal, 14*(3), 179–191.
Porter, M. E. (2008). *On competition*. Harvard Business Press, Boston, USA.
Prahalad, C. K., & Hamel, G. (1995). *Compitiendo por el futuro*. Barcelona: Ariel.
Roman, L. M. (2016). Empresas con base tecnológica e innovación: Emprendimiento rentable. *APLICATEC, 6*(1), 1–5.
Schumpeter, J. (1934). *The theory of economic development*. Cambridge, MA: Harvard University Press.
Scott, B. R. (1989). Competitiveness: Self help for a worsening problem. *Harvard Business Reviews, 67*(4), 115–121.
Sinisterra, O. O. L. (2016). Centers innovation, technological development and entrepreneurship: Case of the Universidad de Panamá. *Hamutay, 2*(2), 40–49.
Teece, D. J. (2007). Explicating dynamic capabilities: The nature and microfoundations of (sustainable) enterprise performance. *Strategic Management Journal, 28*(13), 1319–1350.
Terre-i-Ohme, E. (2002). *Guide for managing innovation: Part 1, diagnosis*. Generalitat de Catalunya, CIDEM, Spain.
Tidd, J., Bessant, J., & Pavitt, K. (2005). *Managing innovation: Integrating technological, managerial organizational change*. New York: Wiley.

Chapter 6
Following the Footprints of SME Competitiveness in a High-Technology Sector

Luís Farinha and Sharmistha Bagchi-Sen

6.1 Theoretical Framework

The globalization of markets brings with it a significant increase of competition and, consequently, an increase in competitiveness across industries (Wacker and Sheu 2006; Aiginger and Vogel 2015). Increasingly, industries need to be attentive to society's problems and needs and ready to develop new solutions for a dynamic world. Innovation, be it product, process, technology, or marketing, is now a driver of competitiveness, becoming the necessary condition for the survival of SME's position in the market (Man et al. 2002; Farinha et al. 2015).

First, it is important to understand the contribution of a strong manufacturing sector to economic prosperity. According to Giffi (2013), the global drivers of manufacturing competitiveness are talent-driven innovation; cost and availability of labour and materials; energy cost and policies; supplier network; local market attractiveness; economic, trade, financial, and tax systems; physical infrastructure; government investments in manufacturing and innovation; legal and regulatory system; market forces; and the healthcare system. However, three factors deserve further attention: (1) talent-driven innovation, that is, the quality and availability of researchers, scientists, and engineers as well as the quality and availability of skilled labour; (2) supplier network, namely, cost competitiveness of local suppliers and the ability of the supply base to innovate products and processes; and (3) investments in

L. Farinha (✉)
Polytechnic Institute of Castelo Branco & NECE at University of Beira Interior, PT, Covilhã, Portugal
e-mail: luis.farinha@ipcb.pt

S. Bagchi-Sen
State University of New York – Buffalo, Buffalo, NY, USA
e-mail: geosbs@buffalo.edu

M. Peris-Ortiz et al. (eds.), *Knowledge, Innovation and Sustainable Development in Organizations*, Innovation, Technology, and Knowledge Management,
https://doi.org/10.1007/978-3-319-74881-8_6

R&D, that is, private and public sector collaboration for long-term sustained investments in science, technology, engineering, and manufacturing.

Second, there is no doubt that innovation guides strategies of differentiation in the market, creating value for the client (e.g. economic value, productivity gains, incremental benefits such as cost efficiency through process innovation) (Ambastha and Momaya 2004; Castellacci 2008). Moreover, networks facilitate an important contribution to increasing firm-level competitiveness (and in turn, regional/national industry level competitiveness), both in terms of collaborative networks for the development of innovative projects, new technologies, and cooperation to access to resources, skills, and/or cost synergies (Polenske 2004; Awazu 2006; Lawton Smith and Bagchi-Sen 2006; Farinha et al. 2016).

Third, clustering and networking as contributors to productive and efficient entrepreneurship and the importance of knowledge and creativity generated from such efforts for sustainable growth are now accepted concepts across stakeholders. Industry clusters, an agglomeration of companies within one particular industry in a specific geographical area, are hotbeds for testing how networks, innovation, and competitiveness come together (Lechner and Dowling 2003; Isaksen 2012; Acs and Amorós 2015; Flåten et al. 2015). In clusters, firms, especially SMEs, engaged in collaborative networks with universities and governments have been shown to demonstrate competitiveness in markets (Etzkowitz and Klofsten 2005; Libaers and Meyer 2011; Park 2012; Felzensztein et al. 2014; Acs and Amorós 2015). This chapter aims to understand the role of such innovation and entrepreneurship networks in SME competitiveness in the Portuguese Engineering and Tooling Cluster. Specific insights on methodologies to understand mechanism of regional cluster dynamics, implications for scholars and practitioners, and future research pathways are also discussed. The chapter is organized into four main sections: theoretical framework, methodology, case study, and conclusions.

6.2 Methodology

6.2.1 Research Position

The analysis presented below aims to contribute to the enrichment of literature in the area of innovation and competitiveness of SMEs through case studies. Case study analysis is a methodology that facilitates researchers to develop grounded research through the construction of interpretative theory, that is, a better understanding of contextualized evidence (Lavine 2000; Yin 2011). Data for understanding the trends in innovation and competitiveness in the Portuguese Engineering and Tooling Cluster are gathered using quantitative and qualitative techniques.

In this study, semi-structured interviews were carried out with managers of cluster-based organizations such as Pool-Net (Portuguese Tooling and Plastics Network) and CENTIMFE (Technological Centre for Mouldmaking, Special

Tooling, and Plastic Industries). In conjunction with the development of the interviews, data about the cluster were collected to facilitate an understanding by the researchers about the study context, that is, the dynamics of innovation and entrepreneurship in that particular geographic cluster. A quantitative analysis was carried out to characterize the business fabric associated with the Portuguese Engineering and Tooling Cluster, from a database from Informa D&B, a company of the CESCE group, specializing in the integral management of commercial risk (Informa D&B 2013). The case study analysis adds a deeper understanding of the reality of the facts such as the types of innovation followed to the initiatives of networks (Trippl and Otto 2009; Isaksen 2012; Balland et al. 2015).

Specific questions investigated in this research are as follows:

What is the role of innovation in SME competitiveness in the Portuguese Engineering and Tooling Cluster?
What are the dynamics of entrepreneurship developed by the cluster with a view to strengthening the competitiveness of the business fabric in the sector?
What is the contribution of university–industry–government (U–I–G) collaborative networks in increasing innovation and competitiveness in the sector?

6.2.2 Unit of Analysis

Portuguese mouldmaking industry started in 1945 by Mr. Aníbal H. Abrantes. The birth of the mouldmaking industry was a consequence of the skills and know-how developed after 1929 in the manufacturing of moulds for glass industry to supply the growing mould needs of Marinha Grande and Oliveira de Azeméis glass industry. The mouldmaking companies and the technical institutions that cooperate together are referred henceforth as the Portuguese Engineering and Tooling Cluster. Composed almost entirely of SMEs, who engage in exports and have favourable balance of trade figures, the Portuguese Engineering and Tooling Industry is of extreme importance to the regional (Portuguese Central Region) and national economy (Camacho 2013; Farinha et al. 2014; Gonçalves et al. 2015).

The 10-year strategic planning[1] of the Portuguese Engineering and Tooling Industry, started in 2008, aims to have recognition in the global market as one of the most advanced value-added sectors in the design and moulding of special tools and high precision metal parts. The mission of the Portuguese mouldmaking industry for plastic injection is to develop and produce moulds by optimizing the process for its customers, ensuring total quality and reducing total costs throughout the product life cycle. In addition, this industry aims to develop and produce special tools and precision machine parts with high value added for the customer, supported by the innovation of manufacturing processes and products. Pool-Net (Portuguese Tooling and Plastics Network) is the entity responsible for the management of the Portuguese

[1] http://www.toolingportugal.com/index.php/en/about-us/vision-and-mission

Engineering and Tooling Cluster. The connection among institutions that support Portuguese mouldmaking industry (CEFAMOL and CENTIMFE), educational and scientific institutions, and mouldmaking companies is facilitated by Pool-Net.

6.3 Case Study

The aim of this case study is to examine the Portuguese Engineering and Tooling Cluster and their growth trajectories. Next, the role of innovation in SME competitiveness is analysed to show the dynamics of entrepreneurship within this cluster.

6.3.1 Engineering and Tooling from Portugal

In the last decades, the Portuguese Engineering and Tooling Cluster has been growing and consolidating its reputation in the global market, driven to a large extent by international demand (Camacho 2013). In the Portuguese case, the mouldmaking industry focuses its activities in the subsector of moulds for plastic and rubber, excluding injection or compression types. The Portuguese economic activity code includes the manufacture of metal moulds. Strongly dependent on the automotive industry, the Portuguese Engineering and Tooling Cluster has been investing in the last decade in strategic sectors such as electronics, energy and environment, electronics, aerospace, and medical devices.

Figure 6.1 shows the location of the Portuguese mouldmaking industry. The two districts, Aveiro and Leiria, are located in the Central Region of Portugal, the second most innovative region of the country, just after Lisbon (Hollanders and Es-Sadki 2017). Thus, these two industrial districts, separated by a little more than 220 km, constitute 80% of the businesses, almost all of them composed of SMEs, which account for about 90% of the total turnover of the sector and employ 87% of the workforce in this type of industry.

The export market of the Portuguese mouldmaking industry has been growing over the last years accounting for more than EUR 600 million in 2013. Portuguese Engineering and Tooling Industry does not depend on the domestic market, with more than 60% of the production exported. Main destinations of Portuguese moulds are Germany, France, Spain, Brazil, Poland, the USA, and Mexico (Camacho 2013; Farinha et al. 2014) (see Table 6.1).

According to Table 6.1, the Portuguese Engineering and Tooling Industry employed more than 8000 workers, producing a gross value added[2] (GVA) of more than 270 million euros in 2013. Approximately 500 companies presented a total

[2]According to the European statistics (Eurostat), Gross Value Added (GVA) is defined as output value at basic prices less intermediate consumption valued at purchasers' prices.

Districts	% Companies	% Turnover	% Employees
Aveiro	**25,0%**	**34,6%**	**35,4%**
Braga	4,2%	2,3%	2,7%
Coimbra	1,3%	2,0%	2,2%
Leiria	**55,3%**	**55,3%**	**51,5%**
Lisboa	4,6%	2,2%	3,2%
Porto	7,8%	3,0%	4,2%
Setúbal	0,4%	0,3%	0,4%

Fig. 6.1 Portuguese Engineering and Tooling Industry. (Source: Based on data from Informa D&B)

Table 6.1 Portuguese Engineering and Tooling Industry

Year	2013
N.° of companies	515
Turnover	689,776,404
Exports	416,345,043
% of exports	60.4%
N° of employees	8019
Gross value added (GVA)	270,288,130
EBITDA% of turnover	12.6%

Source: Informa D&B (2013)

earnings before interest, tax, depreciation, and amortization (EBITDA)[3] in the order of 13% of their turnover in 2013.

More than 70% of companies in the sector have been in existence for more than 10 years, and the percentage of employability is higher than 87% in this group

[3] EBITDA is a measure of a company's operating performance. Essentially, it's a way to evaluate a company's performance without having to factor in financing decisions, accounting decisions, or tax environments.

Table 6.2 Distribution of companies by seniority

Year of foundation	% of companies	Turnover by company	% Employees	Exports (% of turnover)
<2 years	2.52%	65,092	0.3%	73.0%
2–5 years	9.51%	240,368	2.8%	34.9%
6–10 years	17.09%	755,698	9.8%	61.2%
>10 years	70.87%	1,673,016	87.1%	60.7%

Source: Informa D&B (2013)

Table 6.3 Company characterization by size

Size of companies	% of companies	Average of seniority in years	Exports (% of turnover)
Micro	61.00%	12	29.00%
Small	33.00%	20	48.00%
Medium and large	6.00%	29	81.00%

Source: Informa D&B (2013)

(see Table 6.2). However, the exports of the younger companies (e.g. <2 years) represent 73% of their turnover.

Micro enterprises, which employ up to 9 workers, represent more than 60% of the SMEs in the sector (Table 6.3). This group of companies, according to the data of 2013, has up to 12 years of existence and has significant exports (29%) relative to its turnover. The universe of micro and small companies (up to 49 employees) represents 94% of Portuguese Engineering and Tooling Industry, exporting about 77% of the turnover.

The only companies with more than 10 years of existence have on average more than 10 employees (Fig. 6.2). Likewise, the most exporting companies are those that have been around more than 10 years since their foundation. However, from the "Death Valley" (companies with up to 2 years of existence), they have a strong participation in the export market. In terms of turnover and GVA, there is an improvement in these indicators as companies consolidate their presence in the market.

It should also be noted that R&D expenditures are accounted for by companies that have been in business for 6 or more years. This observation reinforces the importance of the establishment of cooperation networks for Research, Development, and Innovation (RD&I) by SMEs with less financial resources (Karna et al. 2013; Lee and Kim 2016; Polenske 2004). Active interactions in RD&I networks of triple and quadruple helix institutional actors such as university–industry–government–civil society (U–I–G–S) can arguably improve innovative capabilities of countries and companies (Farinha et al. 2016; Lawton Smith and Bagchi-Sen 2006, 2012; Lee and Kim 2016; Perianez-Forte and Cervantes 2013). In agreement with this above line of thought (also see Cooke 2010), the Portuguese Engineering and Tooling Cluster assumes itself as a collaborative network based on the triple helix design, that is, it absorbs knowledge spillovers that are easily transformed into portals of knowledge and innovation for the international market. The collaborative dynamics

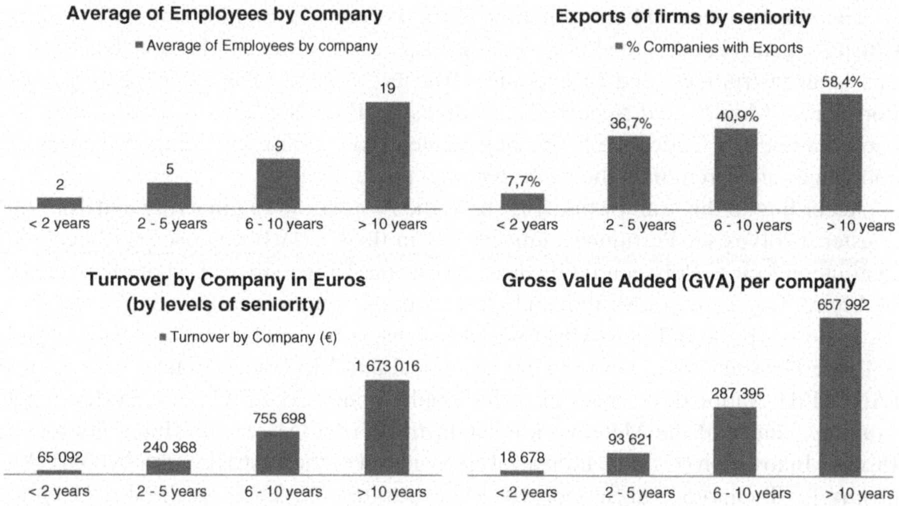

Fig. 6.2 Portuguese mouldmaking industry framework. (Source: Based on data from Informa D&B)

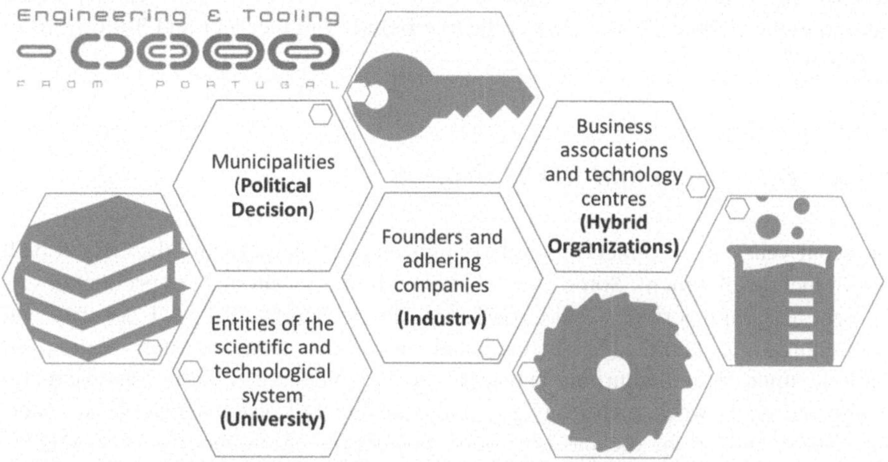

Fig. 6.3 Portuguese Engineering and Tooling triple helix framework

of the Portuguese Engineering and Tooling Cluster is present in the interaction between companies; the scientific and technological system, representative of the capacity of knowledge and technology transfer; municipalities and associations of municipalities, representatives of the regional political decision; and business associations and technology centres, assumed in this configuration (see Fig. 6.3) as hybrid organizations involving additional stakeholders (Leydesdorff and Zawdie 2010; Lawton Smith and Bagchi-Sen 2006).

The management and coordination of the Portuguese Engineering and Tooling Cluster is centralized in the Pool-Net or Portuguese Tooling and Plastics Network – this is an instrument used to operationalize the strategic plan established for the Portuguese Moulds and Special Tools Industry. Pool-Net aims to boost business coordination in a Collective Efficiency Strategy to increase the competitiveness of companies and strengthen their international positioning.

According to this framework (Fig. 6.3), the Portuguese Engineering and Tooling Cluster involves six Portuguese universities in their RD&I processes (Polytechnic Institute of Leiria, Polytechnic Institute of Lisbon, University of Lisbon, University of Aveiro, University of Minho, and University of Coimbra) cooperating with other institutions. In the field of hybrid organizations, active collaboration is developed with the National Association of Metallurgical and Electromechanical Companies (ANEMM), National Association of the Mould Industry (CEFAMOL), Professional Training Centre of the Metalworking Industry (CENFIM), Centre Integration and Process Innovation (CTIM), Industrial Association of the District of Aveiro (AIDA), Centre for Technological Support for Metalworking (CATIM), Technology Centre of Leather Industries (CTIC), Instituto Pedro Nunes (IPN), Institute of Mechanical Engineering and Industrial Management (INEGI), and Hub of Innovation in Polymer Engineering (PIEP). Member companies of the Portuguese Engineering and Tooling Cluster are represented in the domestic market, and especially in the international market, through the collective brand "Engineering and Tooling from Portugal."

6.3.2 Innovation and RD&I Projects

In recent years, the Engineering and Tooling Industry is following the critical path of product development. Some examples of activities developed are product development and engineering, CAD/CAM/CAE, rapid manufacturing, IT security and communication, knowledge management, processes reengineering, high-speed milling, micro-manufacturing, management and planning, quality, environment, health and safety work, metrology, jigs and measurement systems, reverse engineering, plastic injection, die casting tools, powder injection moulding (PIM), bi-injection, reactive injection moulding (RIM), in-mould assembling, and gas injection (see Fig. 6.4).

> *There are a lot of collaboration outside of RD&I projects. There are many companies already using laboratories in universities. This creates some affinity with some aspects at the level of future collaboration ... there is also the placement of fellows from universities in companies with more regularity than a few years ago. (CENTIMFE 2017)*

Portuguese Engineering and Tooling Cluster has three anchor projects and is a further range of complementary projects that together with the anchor projects materialize collective efficiency strategy and the action plan, respectively. The

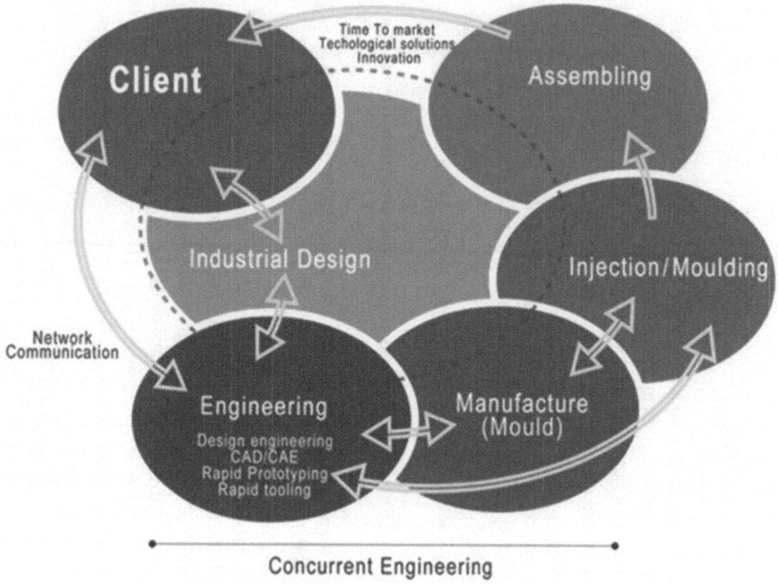

Fig. 6.4 Activities developed in Engineering and Tooling Industry. (Source: Pool-Net)

anchor projects form the basis of cluster development, bringing with it the development of three strategic areas: brand and internationalization (engineering project and tooling from Portugal); technology and technical knowledge associated with new challenges (tooling edge project – sustainable production of high performance); and intervention and intangible factors of competitiveness, specifically with regard to organizational component (ETF project – tooling enterprise of the future) (see Table 6.4).

In order to make more visible the grouping of the projects of Industry Engineering and Tooling, Fig. 6.5 presents typologies within 'product engineering and process' and 'innovation and perspective' – there are still other groups that aim to stimulate industrial and entrepreneurial thinking and cross-cutting development of methodologies to increase competitiveness in SMEs.

6.3.3 *Entrepreneurship and Business Incubator*

Value propositions launched in the marketplace must be based on the needs and expectations of target customer segments (de Andrade Régio et al. 2016; Doganova and Eyquem-Renault 2009; Farinha and Ferrreira 2016; Shafer et al. 2005). In this regard, Pool-Net's vision is that:

Table 6.4 Example of the Portuguese Engineering and Tooling RD&I projects

Project	Area	Sector	Objective
Tooling Edge	Engineering Product and Process I Mechanical Engineering	Industrial Machinery	Develop scientific and technological knowledge, work methodologies and innovative and adapted to Engineering & Tooling Industry, supported by case studies organization. Seeks to enable increasing the overall performance of the industry and add-in their processes and products in a targeted strategy for competitive re-positioning of the cluster value.
Tooling Surface	Engineering Product and Process I Techn. of materials	Health	Studying the factors that affect the flow behaviour of polymeric material during injection moulding.
BigPROTO	Eng. Product and Process I Mechan.Engineering	Aeronautics and Aerospace	Development of materials, equipment, and technologies that enable the manufacture of prototypes or small series of large plastic parts by injection Moulding Reaction of Nylon.
Hibridmolde XXI	Eng. Product and Process I Techn. of materials	Industrial Machinery	Development of an engineering solution for large plastic parts in small series.
SIMI	Eng.Product and Process I Mec. Engineering	Industrial Machinery	Development of moulding tools for the production of multi-material products.
PRODUTECH_PSI	Eng. Product and Process I Automat. and Robotics	Industrial Machinery	New Products for transforming activity: flexible and efficient production systems; energy and environmental efficiency of production systems.
PRODUTECH_PTI	Engineering Product and Process I Automation and Robotics	Industrial Machinery	New and innovative processes at the forefront of manufacturing technologies. Advanced tools for the development of new products and systems. Allow companies to develop new goods and services quickly and efficiently.
X-NANO	Engineering Product and Process I Technology of materials	Industrial Machinery	Study the application of steel shavings, with a ratio nano / micro grain suitable as raw material in two different ways: framing of various products by injection of powders; and production of hybrid materials from sheet micrometer thickness after dynamic consolidation and rolling followed by micro-injection polymer.
CoolMOULD	Engineering Product and Process I Technology of materials	Industrial Machinery	Studying systems and solutions that improve the thermal and energy performance mould to produce high quality parts at a lower cost, allowing the development of a new product, that is, the mould in a more efficient thermal and energy terms.

METALMORPHOSUS	Eng. Product and Process | Techn. of materials	Industrial Machinery	Study of EMF technology for joining parts in composite materials and metals to obtain hybrid components. The role of Centimfe is the coordination and sharing of R&D at the level of injecting composites.
MicroHANDLING	Eng. Product and Process | Moulds – Handling of parts	Industrial Machinery	Development of a new value chain directed to the provision of moulds for micro parts, the handling systems, and micro-assembly.
BestCRANK	Engineering Product and Process | Technology of materials	Industrial Machinery	Intends to develop a new product with innovative features on the bicycle accessory market. The project will study different materials, geometries and methodologies, thus providing knowledge to develop a design that will maximize resistance that will stand by a high strength / weight / exclusivity.
RITECA II	European project	Research Network	Strengthening the cooperation network between technological and research centres, in Centro and Alentejo Portuguese Regions and Spanish Extremadura.
GAPI Horizon	Innovation and Prospective	Industrial Property	Development activities related to the virtual network, interconnection and communication with IDI RITECA partners.
Platform for Knowledge Transfer	Innovation and Prospective	Sharing and Knowledge Transfer Network	Create a network of sharing and knowledge transfer in a logical approach to the "production of knowledge" and the business community.
Innovation Engineering & Tooling Platform	Innovation and Prospective	Engineering & Tooling Innovation Network	The Network Engineering & Tooling Innovation constitutes a knowledge network that will support and complement the activities undertaken by the Portuguese E&T Cluster under the collective efficiency strategies in order to increase the representativeness of the cluster internationally.
Pense Indústria	Sensitizing Youth for Industry	Industry	Motivate and sensitize young students of the Elementary and Secondary Education so they opt for future professional careers related to the business activity of the industrial base.
ETF	Transverse	Industry	The ETF Project – "Tooling Company of the Future" aims to develop methodologies to support companies in the industry Engineering & Tooling, aiming to broaden the competitive basis, that is, repositioning its offering to strategic sectors.

(continued)

Table 6.4 (continued)

Project	Area	Sector	Objective
ReMOULD	Transverse	Training	Project training for the retraining of older workers in the field of injection moulding.
LINK2FP7	Innovation and Prospective	European networks	The project aims at integration of stakeholders, including companies in the Engineering & Tooling Cluster, in the context of European RDI networks (in European programs).
GESTOOLING	Innovation and Prospective	Industrial Property	Management and Activities Network. Registration, management and brand promotion "Engineering and Tooling from Portugal." Stimulation of Partnerships and Cooperation Networks. Dissemination and Knowledge Sharing.
Btm – Branding to Market	Innovation and Prospective	Industrial Property	The "BTM – Branding to Market" project aims to promote and consolidate the brand "Engineering and Tooling from Portugal", as a collective, distinguishing and differentiating brand of Industry Engineering & Tooling at national and international level, fostering competitiveness and strengthening the its position in international markets.
GESTOOLING 2014	Innovation and Prospective	Industrial Property	Management and Activities Network. Stimulation of Partnerships and Market Opportunities. Dissemination and Sharing of Information and Knowledge.
Btm 2015 Branding to Market	Innovation and Prospective	Industrial Property	Aims to continue the international campaign promoting the brand and Cluster "Engineering and Tooling from Portugal", fostering competitiveness and strengthening its position in international markets.
Wiintech	Innovation and Prospective	European networks	The project Wiitech develops a common international strategy for cooperation between European clusters, seeking to build interclusters international partnerships, with a focus on clean technologies. This cooperation aims in particular to optimize partnerships and protocols with other Clusters of Japan, USA, Brazil and India.
In-Tooling	Innovation and Prospective	Transverse	This project aims to enhance the intelligence of the cluster through the promotion of a set of strategic studies for the industry, and working for Tooling Technology Roadmap in perspective Horizon 2020.

Source: Pool-Net

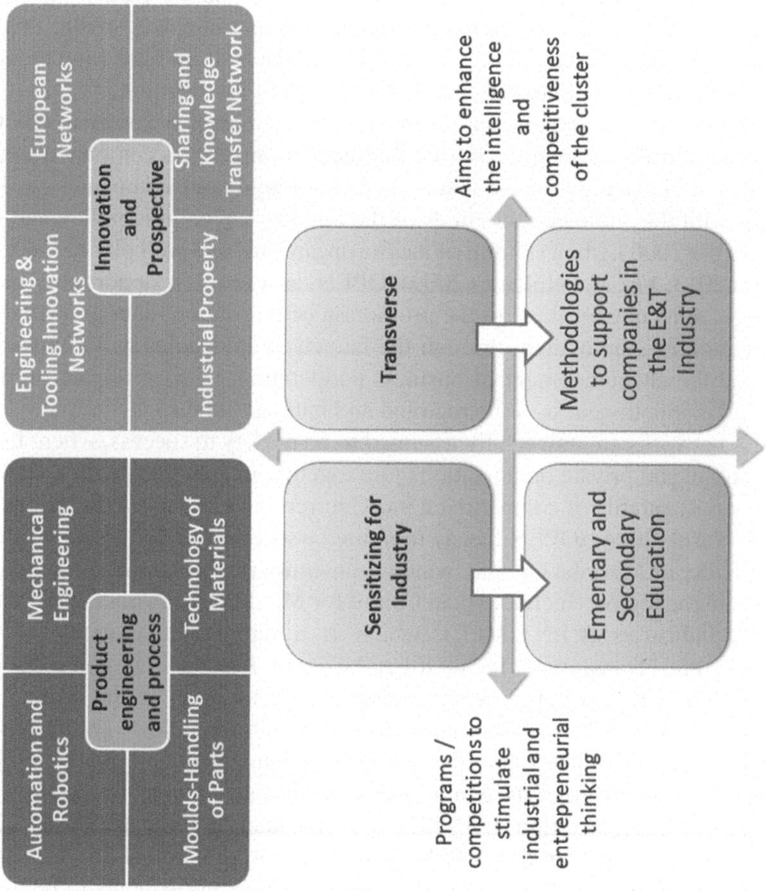

Fig. 6.5 Projects by nature

In the automotive industry, the customer not only wants the moulds, they want the dashboard of the car ... the market requirement sets the corporate strategy, extending its value chain from design, to prototyping, engineering, production, marketing and solution delivery.

Under the motto 'to undertake innovating', the OPEN project, which is now open to entrepreneurs and to society in general, corresponds to the concretization of a strongly mature idea, which began in 1997 with the project Territorial Pact for Employment of Marinha Grande (District of Leiria). At that time, the need to launch an initiative that would facilitate strong entrepreneurship and the capacity to innovate and acknowledge the entrepreneurial dynamism that is characteristic of this region was recognized. This initiative supported the interests and the competences of a group of promoters with experience in their areas of intervention.

Given that knowledge and innovation are recognized as decisive elements for the competitiveness of the economy and that the incubation process can play a major role, the idea of launching an incubator of technology-based companies, in full compliance with the strategic guidelines of the country's economic policy and the objectives of the 2000 Lisbon summit of the European Union, is well placed (Isaksen and Nilsson 2013; Vaz and Nijkamp 2009). OPEN, as a business opportunity incubation centre, aims to contribute to the promotion of innovation, entrepreneurship, and the creation of employment through the launch of companies with innovative concepts and the encouragement of business cooperation, with an impact on productivity and competitiveness at the regional and national levels.

Business networks are universally assumed to be the key to success, where firms and other public and private organizations join together in networks, with a view to achieving new standards of competitiveness (Guerrero and Urbano 2017; Kingsley et al. 2011; Turok 2004). OPEN has as founding associates the following entities: the Municipality of Marinha Grande, which is an entity representative of the political decision in the region; Technological Centre for Mouldmaking, Special Tooling, and Plastic Industries (CENTIMFE), which is a representative entity of the Development and Transfer of Technology; National Association of the Mould Industry (CEFAMOL), which is a representative of industries in the sector; the National Association of Young Entrepreneurs (ANJE) in the field of entrepreneurship; and INOVA, which is a supportive entity for organization and quality. OPEN is also part of a partnership of the Municipality of Marinha Grande, the Agency for Competitiveness and Innovation (IAPMEI), Banco Santander Totta, and GARVAL – Venture Capital Society, for the coordination and management of the FINICIA Marinha Grande Fund, which aims to stimulate and guide investments to be made for micro and small businesses existing or in the creation phase in the Municipality of Marinha Grande, with activity or business project of local relevance. Financing up to the amount of € 45,000 is provided.

The relationships between businesses and the scientific and technological system have existed for many years, and companies have been working collaboratively for a long time. Moreover, business missions and collective participation in fairs have been happening for many years. The recognition of competitiveness in E&T came to help consolidate and formalize this collaborative relationship. (Pool-Net Association)

OPEN target groups are innovative projects in both industrial and service areas – they are young people, highly qualified at the beginning of their career and with an entrepreneurial spirit; researchers and other professionals with high potential for bringing together knowledge, experience, and academic skills; and companies whose reconversion or specialization in emerging technological areas boosts the creation of new entrepreneurial projects and spin-offs.

The main support services available are support for the constitution and legalization of companies – this is done through the following activities: conducting training, seminars, workshops, and conferences, supporting the dissemination of developed technologies, providing advice on the most appropriate incentive programs for projects, supporting access to risk capital and seed capital, helping in the preparation of business plans and strategic plans, participating in national and foreign cooperation networks, supporting the implementation and management of quality assurance systems, and supporting the implementation and management of systems of hygiene, health, and safety in the workplace. Some of the completed and ongoing projects are described in Table 6.5.

Table 6.5 RD&I projects developed with support from OPEN

Project	Description
CEMICRO – Micro Manufacturing Competence Network	The CEMICRO Project has as its mission the cooperation between entities of the National Scientific and Technological System and the business fabric, capable of inducing the development of skills and partnerships to support the development and industrialization of microcomponents and microdevices
Clusterplast – Intercluster initiative to target the future challenges for the European polymer converting industry	The Clusterplast project is a European intercluster initiative – bringing together clusters, in the fields of the value chain of the Moulds and Plastics Industries of France, Portugal, Austria, Spain, Italy, and the Czech Republic – which aims, through the definition of a Joint Action Plan and co-promotion of intercluster initiatives, to promote bases for the development of these clusters and ultimately for sustainability and to respond to the future challenges of the European industry
IDT Explore	The IDT Explore Project – Creative and Innovative Community is focused on the scope of Measure 1.5 of the Operational Program of the Central Region – Triangular Projects and is promoted by the IDT Network
Community Initiative Program INTERREG III B SUDOE – Tecnoemprende Project	The Tecnoemprende project is born with the vocation of being the first measure of an interregional nature designed to facilitate the creation of technology-based companies in South-West Europe (SUDOE)
Project 'START UP – A RIERC Initiative' – Central Region Incubation and Entrepreneurship Network	OPEN is revitalizing the START UP project – a Centre for Incubation and Entrepreneurship RIERC Initiative. This project was recently approved within the scope of the Support System for Business Reception and Logistics Areas

(continued)

Table 6.5 (continued)

Project	Description
TII Project – Technology, Innovation, and Initiative	Project TII – Technology, Innovation, and Initiative promoted by OPEN in partnership with Technological Centre of the Textile and Clothing Industry (CITEVE) and MICROSOFT
Project IC|16 – Centro Incubation 2016	The application submitted by RIERC to Portugal 2020 was recently approved under the measure Incentive System for Collective Actions (SIAC), 'Promotion of Entrepreneurship', in the following typology: training of entrepreneurial initiatives and the creation of new companies

Source: OPEN Incubator

6.4 Conclusions

The globalization of markets brings with it a strong increase of competition in the industry, in particular in the Engineering and Tooling Sector. Innovation is expected to improve industrial competitiveness in the globalized market (Acs and Amorós 2015; Acs et al. 2001; Carlsson and Mudambi 2003; Polenske 2004; Stejskal and Hajek 2012). The prosperity of SMEs in mouldmaking is based on its capacity to innovate and offer differentiating products to the market and provide solutions for problems in its target market. Creativity, clustering, and collaborative networks for RD&I are key for sustainable growth and competitiveness of SMEs (Camacho 2013).

The Portuguese mouldmaking industry focuses its activities in the subsector of moulds for plastic and rubber (high-precision metal moulds). To reduce the heavy dependence on the automotive industry, the Portuguese Engineering and Tooling Cluster has been investing in strategic sectors such as electronics, energy and environment, electronics, aerospace, and medical devices sector. Composed mostly of micro and small companies, the Portuguese Engineering and Tooling Sector employs more than 8000 workers, with a turnover in the order of 690 million euros, mostly destined for the export market.

Based on the quadruple helix model (Leydesdorff 2012; Prainsack 2012), the Portuguese Engineering and Tooling Cluster is present in the interaction between companies and the scientific and technological system. It represents the capacity of knowledge and technology transfer in municipalities and associations of municipalities, who are representatives of the regional political decision. Other stakeholders are business associations and technology centres and the hybrid organizations. In the context of RD&I, the Engineering and Tooling Cluster has become competitive in the market, investing heavily in projects aimed at "product and process engineering" as well as "innovation".

In the field of entrepreneurship, the cluster has OPEN, which is a business opportunity incubation centre, that aims to promote innovation, entrepreneurship, and the creation of innovative companies in the mouldmaking sector, encouraging cooperation among various stakeholders connected to the business itself (Z. Acs and Amorós 2015).

Acknowledgments We would like to thank the employees of Pool-Net (Portuguese Tooling and Plastics Network), CENTIMFE, and OPEN Incubator for their valuable input. We would like to express our sincere thanks to Informa D&B for the data on the sector.

References

Acs, Z., & Amorós, J. (2015). Entrepreneurship and development: The role of clusters. *Small Business Economics*. Springer Berlin Heidelberg, *39*(3), 561–574. https://doi.org/10.1007/s11187-015-9643-3.

Acs, Z. J., Morck, R. K., & Yeung, B. (2001). Entrepreneurship , globalization , and public policy. *Journal of International Management, 7*, 235–251.

Aiginger, K., & Vogel, J. (2015). Competitiveness : From a misleading concept to a strategy supporting beyond GDP goals. *Competitiveness Review, 25*(5), 497–523. https://doi.org/10.1108/CR-06-2015-0052.

Ambastha, A., & Momaya, K. (2004). Competitiveness of firms-review of theory, frameworks and models. *Singapore Management Review, 26*(1), 45–61. https://doi.org/10.1002/ccd.10430.

Awazu, Y. (2006). Managing technology alliances: The case for knowledge management. *International Journal of Information Management, 26*(6), 484–493. https://doi.org/10.1016/j.ijinfomgt.2006.07.005.

Balland, P.-A., Boschma, R., & Frenken, K. (2015). Proximity and innovation: From statics to dynamics. *Regional Studies, 49*(6), 907–920. https://doi.org/10.1080/00343404.2014.883598.

Camacho, J. F. (2013). In I. World (Ed.), *ISTMA statistical year book: On tools, dies and moulds industry – International trade and manufacturing* (5th ed.). Lisbon: ISTMA World.

Carlsson, B., & Mudambi, R. (2003). Globalization, entrepreneurship, and public policy: A systems view. *Industry & Innovation, 10*(1), 103–116. https://doi.org/10.1080/1366271032000068122.

Castellacci, F. (2008). Innovation and the competitiveness of industries: Comparing the mainstream and the evolutionary approaches. *Technological Forecasting and Social Change, 75*(7), 984–1006. https://doi.org/10.1016/j.techfore.2007.09.002.

Centimfe. (2017). Centimfe–Market, services and projects. Retrieved by October 15, from http://www.centimfe.com.

Cooke, P. (2010). Regional innovation systems: Development opportunities from the "green turn". *Technology Analysis & Strategic Management, 22*(March 2013), 831–844. https://doi.org/10.1080/09537325.2010.511156.

de Andrade Régio, M. M., et al. (2016). Forecasting the disruptive skillset alignment induced by the forthcoming industrial revolution. *Romanian Review Precision Mechanics, Optics & Mecatronics, 49*(1), 24–29 Available at: http://search.ebscohost.com/login.aspx?direct=true&site=eds-live&db=a9h&AN=117202208.

Doganova, L., & Eyquem-Renault, M. (2009). What do business models do? *Research Policy, 38*(10), 1559–1570. https://doi.org/10.1016/j.respol.2009.08.002.

Etzkowitz, H., & Klofsten, M. (2005). The innovating region: Toward a theory of knowledge-based regional development. *R and D Management, 35*(3), 243–255. https://doi.org/10.1111/j.1467-9310.2005.00387.x.

Farinha, L., & Ferrreira, J. J. (2016). Measuring innovative and entrepreneurial networks performance. *International Journal of Social, Behavioral, Educational, Economic, Business and Industrial Engineering, 10*(8), 2673–2682.

Farinha, L., Ferreira, J. J., & Gouveia, J. B. (2014). Innovation and competitiveness: A high-tech cluster approach. *Romanian Review Precision Mechanics, Optics & Mechatronics, 45*, 41–48.

Farinha, L., Gouveia, J. B., & Nunes, S. (2015). Drivers of global competitiveness as economic growth. In L. Farinha et al. (Eds.), *Handbook of research on global competitive advantage through innovation and entrepreneurship* (1st ed., pp. 13–27). Hershey: IGI Global. https://doi.org/10.4018/978-1-4666-8348-8.ch002.

Farinha, L., Ferreira, J., & Gouveia, B. (2016). Networks of innovation and competitiveness: A triple helix case study. *Journal of the Knowledge Economy, 7*(1), 259–275. https://doi.org/10.1007/s13132-014-0218-3.

Felzensztein, C., et al. (2014). International marketing strategies in industrial clusters: Insights from the southern hemisphere. *Journal of Business Research.*, Elsevier Inc., *67*(5), 837–846. https://doi.org/10.1016/j.jbusres.2013.07.002.

Flåten, B.-T., Isaksen, A., & Karlsen, J. (2015). Competitive firms in thin regions in Norway: The importance of workplace learning. *Norsk Geografisk Tidsskrift – Norwegian Journal of Geography, 69*(2), 102–111. https://doi.org/10.1080/00291951.2015.1016875.

Giffi, C. A. (2013). *Global manufacturing competitiveness index* (p. 40). Washington: Deloitte.

Gonçalves, V., et al. (2015). Twenty years after the porter report for Portugal. *Competitiveness Review, 25*(5), 540–554. https://doi.org/10.1108/CR-06-2015-0059.

Guerrero, M., & Urbano, D. (2017). The impact of triple helix agents on entrepreneurial innovations' performance: An inside look at enterprises located in an emerging economy. *Technological Forecasting and Social Change*. Elsevier Inc., *119*, 294–309. https://doi.org/10.1016/j.techfore.2016.06.015.

Hollanders, H., & Es-Sadki, N. (2017). In E. Commission (Ed.), *Regional innovation scoreboard 2017*. Washington: European Commission. https://doi.org/10.2873/469382.

Informa D&B. (2013). Informa D&B Database - Portuguese industry of moulds. Lisboa.

Isaksen, A. (2012). Regional clusters and competitiveness : The Norwegian case. *European Planning Studies, 5*(2), 65–76. https://doi.org/10.1080/09654319708720384.

Isaksen, A., & Nilsson, M. (2013). Combined innovation policy: Linking scientific and practical knowledge in innovation systems. *European Planning Studies, 21*(12), 1919–1936. https://doi.org/10.1080/09654313.2012.722966.

Karna, A., Täube, F., & Sonderegger, P. (2013). Evolution of innovation networks across geographical and organizational boundaries: A study of R&D subsidiaries in the Bangalore IT cluster. *European Management Review, 10*(4), 211–226. https://doi.org/10.1111/emre.12017.

Kingsley, G., Malecki, E. J., & Malecki, J. (2011). Networking for competitiveness. *Small Business Economics, 23*(1), 71–84. https://doi.org/10.1023/B:SBEJ.0000026022.08180.b7.

Lavine, S. (2000). Quantification and ontology. *Epistemology & Philosophy of Science, 3*, 1–43. https://doi.org/10.1023/A:100525742.

Lawton Smith, H., & Bagchi-Sen, S. (2006). University–industry interactions: The case of the UK biotech industry. *Industry & Innovation, 13*(4), 371–392. https://doi.org/10.1080/13662710601032697.

Lawton Smith, H., & Bagchi-Sen, S. (2012). The research university, entrepreneurship and regional development: Research propositions and current evidence. *Entrepreneurship & Regional Development, 24*(5–6), 383–404. https://doi.org/10.1080/08985626.2011.592547.

Lechner, C., & Dowling, M. (2003). Firm networks: External relationships as sources for the growth and competitiveness of entrepreneurial firms. *Entrepreneurship & Regional Development, 15*(1), 1–26. https://doi.org/10.1080/08985620210159220.

Lee, Y. H., & Kim, Y. J. (2016). Analyzing interaction in R&D networks using the triple helix method: Evidence from industrial R&D programs in Korean government. *Technological Forecasting and Social Change.*, Elsevier Inc., *110*, 93–105. https://doi.org/10.1016/j.techfore.2015.10.017.

Leydesdorff, L. (2012). The triple helix, quadruple helix, …, and an N-tuple of helices: Explanatory models for analyzing the knowledge-based economy? *Journal of the Knowledge Economy, 3*(1), 25–35. https://doi.org/10.1007/s13132-011-0049-4.

Leydesdorff, L., & Zawdie, G. (2010). The triple helix perspective of innovation systems. *Technology Analysis & Strategic Management, 22*(7), 789–804. https://doi.org/10.1080/09537325.2010.511142.

Libaers, D., & Meyer, M. (2011). Highly innovative small technology firms, industrial clusters and firm internationalization. *Research Policy*. Elsevier B.V., *40*(10), 1426–1437. https://doi.org/10.1016/j.respol.2011.06.005.

Man, T. W. Y., Lau, T., & Chan, K. F. (2002). The competitiveness of small and medium enterprises A conceptualization with focus on entrepreneurial competencies. *Journal of Business Venturing, 17*(June 1998), 123–142.

Ortega Argiles, R., McCann, P., Perianez-Forte, I., Cervantes, M., Larosse, J., & Sanchez, L. (2013). Innovationdriven growth in regions: The role of smart specialisation. (OECD Science, Technology and Industry Policy Papers; Vol. 12). Paris: OECD/ODCE.

Park, S. C. (2012). Competitiveness of east Asian science cities: Discourse on their status as global or local innovative clusters. *AI and Society, 27*(4), 451–464. https://doi.org/10.1007/s00146-011-0367-x.

Perianez-Forte, I., & Cervantes, M. (2013). *Innovation-driven growth in regions : The role of smart specialisation (Preliminary Version)*, 202.

Polenske, K. R. (2004). Competition, collaboration and cooperation: An uneasy triangle in networks of firms and regions. *Regional Studies, 389*(November), 1029–1043. https://doi.org/10.1080/0034340042000292629.

Prainsack, B. (2012). Elias G. Carayannis and David F. J. Campbell, mode 3 knowledge production in quadruple helix innovation systems: 21st-century democracy, innovation, and entrepreneurship for development. *Minerva, 50*(1), 139–142. https://doi.org/10.1007/s11024-012-9194-6.

Shafer, S. M., Smith, H. J., & Linder, J. C. (2005). The power of business models. *Business Horizons, 48*(3), 199–207. https://doi.org/10.1016/j.bushor.2004.10.014.

Stejskal, J., & Hajek, P. (2012). Competitive advantage analysis: A novel method for industrial clusters identification. *Journal of Business Economics and Management, 13*(2), 344–365. https://doi.org/10.3846/16111699.2011.620154.

Trippl, M., & Otto, A. (2009). How to turn the fate of old industrial areas: A comparison of cluster-based renewal processes in Styria and the Saarland. *Environment and Planning A, 41*(5), 1217–1233. https://doi.org/10.1068/a4129.

Turok, I. (2004). Cities, regions and competitiveness. *Regional Studies, 38*(9), 1069–1083. https://doi.org/10.1080/0034340042000292647.

Vaz, T. D. N., & Nijkamp, P. (2009). Knowledge and innovation: The strings between global and local dimensions of sustainable growth. *Entrepreneurship & Regional Development, 21*(4), 441–455. https://doi.org/10.1080/08985620903020094.

Wacker, J. G., & Sheu, C. (2006). Effectiveness of manufacturing planning and controlsystems on manufacturing competitiveness: Evidence from global manufacturing data. *International Journal of Production Research, 44*(5), 1015–1036. https://doi.org/10.1080/00207540500268681.

Yin, R. K. (2011). *Qualitative-research-from-start-to-finish*. New York: The Guilford Press.

Chapter 7
The Integration of the Supply Chain as a Dynamic Capability for Sustainability: The Case of an Innovative Organic Company

Maria D. Moreno-Luzon, Juan P. Escorcia-Caballero, and Odette Chams-Anturi

7.1 Introduction

Companies are currently facing a scenario of ever-greater global competitiveness, which has forced innovation to become more frequent and greater in scope. Consequently, in order to be competitive, companies must be able to develop innovation in the shortest possible time spans and at the lowest cost while ensuring the sustainability of their supplies. These objectives are difficult to achieve when organizations work alone due to limited internal resources. Therefore, competition in the global market no longer centers on business to business relationships but instead resides in competition between supply chains (Näslund and Hulthen 2012).

The integration of supply chain processes is considered a key factor in supply chain management (SCM). Lambert and Cooper (1998) argued that the purpose of integration is to improve the efficiency and effectiveness of the processes of supply chain (SC) members. Anderson and Katz (1998) postulated that integration enables the creation and coordination of processes through SC in such a way that most competitors cannot easily match the advantages obtained. One of the key dimensions of supply chain integration (SCI) is integration with suppliers, which refers to the degree to which a firm teams up with its suppliers to structure inter-organizational strategies, practices, and processes in a collaborative and synchronized way (Stank and Keller 2001). This type of integration requires firms to look beyond their organizational boundaries and evaluate how the resources and capabilities of their suppliers can adapt to their needs.

M. D. Moreno-Luzon · J. P. Escorcia-Caballero · O. Chams-Anturi(✉)
Department of Management, Faculty of Economics, University of Valencia, Valencia, Spain

Business School, Universidad del Norte, Barranquilla, Colombia
e-mail: ochamsan@alumni.uv.es; escajuan@alumni.uv.es; maria.moreno@uv.es

M. Peris-Ortiz et al. (eds.), *Knowledge, Innovation and Sustainable Development in Organizations*, Innovation, Technology, and Knowledge Management,
https://doi.org/10.1007/978-3-319-74881-8_7

In recent years, the relationship between supplier integration and organizational innovation has been studied using different theoretical frameworks. In general, the literature suggests that integration with suppliers is an important factor for businesses because it helps them to obtain information and knowledge and to develop capabilities that firms do not have but need to implement innovations (Lau 2014). However, most of the research that has studied this relationship focuses on the benefits that the firm obtains through integration with its suppliers, forgetting that the activities carried out in supplier integration should also consider how firms help their suppliers to understand changes in the environment. In addition, much of the research does not distinguish between the degree of innovation to be developed, which has meant that the effects of supplier integration on radical innovation have not been fully understood, and there are even doubts as to whether a high degree of supplier integration actually facilitates its development.

Accordingly, the objective of the present research is to understand the role of supplier integration on the development of radical innovation, with a particular focus on how this type of integration facilitates the adaptation of the firm and its suppliers to the new requirements that emerge out of radical innovation. To do this, we used the dynamic capabilities' framework, as this allows us to understand how the routines involved in supplier integration make it easier for firms to address changes in their environment. In addition, we analyzed this connection through an explorative case study in an innovative organic agro-food firm.

This chapter is organized as follows: the next section introduces and defines supplier integration and explains its relationship with innovation. The second section conceptualizes supplier integration as a set of routines. The third section conceptualizes supplier integration as a dynamic capacity and highlights the role of supplier integration routines in promoting company adaptation. The fourth section shows the results of an exploratory case study in an organic agro-food firm, providing an insight into the role of routines in making supplier integration a dynamic capability and its connection to radical innovation. The last section discusses the results within this theoretical framework and their implications for future research.

7.2 Supplier Integration and its Role on Innovation

In general, there is evidence in SCM literature that supplier integration makes it easier for a supplier to anticipate the needs of the firm, managing to adequately satisfy changes in requirements. Specifically, in the context of innovation, it is recognized that the probability of an innovation succeeding in the market is highly correlated with supplier understanding of a buyer's needs (Teece 2007). Therefore, the importance of supplier integration to innovation is based on the fact that it helps firms to establish a mutual understanding with their suppliers, allowing them to share knowledge, improve processes, and jointly develop new products (Lau 2011; Petersen et al. 2005; Ragatz and Handfield 1997; Thomas and Griffin 1996).

For example, when Petersen et al. (2005) studied the relationship between the key variables of supplier integration and the development of new products, they found that supplier participation in the establishment of technical objectives and of business metrics was associated with the project team's effectiveness during the process of the new product's development. Soosay et al. (2008) examined the strategies adopted by different firms to promote collaborative relationships with their suppliers that enabled the development of innovation capacities. The results obtained indicated that activities such as joint planning, knowledge exchange, and synchronization of activities facilitate continuous innovation in firms. Moreover, He et al. (2014) studied the direct and indirect impact of supplier integration on the performance of new products. The results indicated that supplier integration has both a direct and indirect effect on the development of new products, thus supporting its importance for innovation.

7.3 The Role of Supplier Integration on Radical Innovation

Radical innovations are defined as fundamental changes in new products that represent revolutionary changes in products and process technologies (Song and Di Benedetto 2008), offering substantial new benefits to customers (Chandy and Tellis 1998). By association, these innovations involve greater risks for the company due to their high degree of uncertainty and the large amount of new knowledge required (Leifer 2000). Radical innovations are important because they let organizations to adapt or even create disruptive changes in its environment that ensure its long-term survival (McDermott and O'Connor 2002).

In SCI literature, there are some discrepancies as to the role of supplier integration on radical innovation. Some research suggests that low levels of supplier integration are required when innovations involve a high level of new technology or product novelty (Eisenhardt and Tabrizi 1995; Luo 2002; Swink 1999) because current suppliers use resources and capabilities that are not suitable for the firm's new requirements (Lau 2014). Therefore, supplier integration is more closely linked to the improvement of organizational productivity and efficiency (Bonaccorsi and Lipparini 1994; Flynn et al. 2010; Mentzer et al. 2000; Walter and Gemünden 2000), which are aspects that have been recognized as ensuring the survival of companies in stable environments but not in highly dynamic environments. For example, Lau (2014) studied the effect of supplier integration on the development of new products. In a sample of 251 firms from different industries in Hong Kong, empirical evidence supporting the positive effect of supplier integration on the performance of new products was found. However, there was also evidence that firms that faced situations of high uncertainty tended to pursue low levels of supplier integration since this meant that firms could swap suppliers to meet new market and technological changes.

On the other hand, contrary to what has been mentioned above, other research has argued that supplier integration facilitates the development of radical innovation, because it helps companies to explore external resources and knowledge (Wong et al. 2013), letting the search for new and better ways of serving customers (Ittner and Larcker 1997). It is recognized in the literature that it is difficult for firms to develop innovations that require a high degree of new knowledge on their own, and preferably supply chain level efforts are required, which include joint efforts with suppliers (Lin et al. 2010). For example, Song and Di Benedetto (2008) provided empirical evidence of the importance of suppliers to radical innovation. The results obtained from 173 projects classified as radical innovations indicate that the greater the participation of major suppliers, the greater the benefits obtained from the new products. In general, this research highlights that supplier integration facilitates the development of radical innovation as it offers technological changes and reduces competency traps (Ragatz et al. 2002). In addition, suppliers can offer innovative ideas to manufacturers and co-design new products, offering more cost-effective design options, an appropriate selection of components and technologies, and access to resources beyond the manufacturer's boundaries (He et al. 2014; Narasimhan and Das 1999).

Based on our literature review, we believe that the contradictory results obtained from the role of supplier integration on radical innovation are because most research has studied this connection from a unidirectional point of view. That is, it is based on the ability of the firms to assimilate external knowledge that originated from integration activities with its suppliers for the purpose of achieving radical innovation. Nevertheless, less is understood about the flip side of the coin, which refers to the need for firms to provide knowledge to its suppliers and to help them assimilate it, so that their innovations can be implemented. Therefore, in order to understand how integration with suppliers can facilitate the development of radical innovation, this chapter aims to conceptualize supplier integration as a dynamic capability. We start from perspectives that conceptualize supplier integration as an organizational capacity, identifying the routines that make it up and then classifying them according to the dynamic capabilities of sensing, seizing, and transformation proposed by Teece (2007) and utilized by Vanpoucke et al. (2014) within a supplier integration framework.

7.4 Supplier Integration as an Organizational Capability

Before understanding supplier integration as a dynamic capability, we began by conceptualizing supplier integration as an operational capability. We used the framework developed by Peng et al. (2008), who argued that the capabilities of an organization are built through the identification, development, and integration of organizational routines (Peng et al. 2008). Organizational routines are described as the way things are done (Teece et al. 1997) and are repetitive and recognizable patterns of interdependent actions carried out by multiple actors (Feldman and Pentland 2003).

Recent research suggests the use of a routine-based approach to understand how the firm manages its suppliers (Day et al. 2015), since according to Feldman and Pentland (2003), routines have a collective nature because they involve multiple actors, thus meaning that their execution implies actors in different locations. This means routines can be distributed throughout the organization or throughout the supply chain, and their actors can belong to different departments of the organization or even be part of different organizations connected by the interactions between them. Therefore, it is important to study the capabilities of the supply chain through its underlying routines and particularly those used for the integration of resources and competencies (Peng et al. 2008, p.744).

From this perspective, we can define the supplier integration capability as a set of routines to carry out inter-organizational collaboration and cooperation activities, in order to improve the flow of information, material, finance, and decision-making issues between the organization and its suppliers. Supplier integration includes activities such as joint product development, information-sharing, and process coordination (Day et al. 2015; Ragatz et al. 2002; Zhao et al. 2008). Based on our literature review and the analysis of different measurement scales used to measure the levels of supplier integration in the context of the SC (Chen and Paulraj 2004; Day et al. 2015; Flynn and Zhao 2014; Frohlich and Westbrook 2001; Huo et al. 2014; Lai et al. 2012; Morgan and Monczka 1996; Narasimhan and Kim 2002; Petersen et al. 2005; Swink and Nair 2007; Wang et al. 2016), we identified activities and patterns that were grouped into four main routines, which we will call assessment routines, process coordination routines, information-sharing routines, and joint product routines.

7.4.1 Assessment Routines

Assessment routines serve to select the potential suppliers with which the company can integrate and monitor their current performance in the pursuit of integration, thus maintaining a continuous evaluation system that avoids opportunism and identifies points of improvement and new opportunities (Chen and Paulraj 2004; Day et al. 2015). Therefore, in order to achieve high levels of supplier integration, it is important to assess their skills in terms of quality, delivery, capabilities, and process compatibility, among others (Song and Di Benedetto 2008). For example, Petersen et al. (2005) suggested that to achieve successful integration, a detailed supplier assessment should be carried out, including aspects such as a selection of the appropriate supplier, complementarity of capabilities, cultural aspects, and integration processes.

7.4.2 Information-Sharing Routines

Information-sharing routines let firms to share what happens in the supply chain. Generally, they involve different types of data and knowledge such as inventory levels, demand forecasts, production plans, product traceability, and the technical characteristics of products (Lau 2014; Tempelaar and Van Den Vrande 2012; Vanpoucke et al. 2014). Therefore, to obtain high levels of integration, the information exchange must be frequent and bidirectional (Müller and Gaudig 2011). Information-sharing can be done through meetings and via telephone, mail, and the interchange of electronic data, enabling better decision-making in the supply chain. For example, information-sharing is recognized as being essential for the development of new products, because it enables needs to be identified and development processes to be streamlined (Monczka 2000; Ragatz et al. 2002).

7.4.3 Process Coordination Routines

Process coordination routines let firms to coordinate and structure its relationship with suppliers, allowing future problems to be understood, external knowledge to be incorporated into current planning decisions, and proactive management of opportunities and threats to physical flows (Vanpoucke et al. 2014). Therefore, in order to improve levels of supplier integration, the company must work with their suppliers to plan activities and solve problems, allowing them to co-align their operations and processes, and to increase trust and commitment (Nyaga et al. 2010). This helps to reduce redundant efforts between the company and its suppliers and facilitates joint efforts to lower costs, improve quality, and leverage resources (Rosenzweig et al. 2003; Swink and Song 2007; Wong et al. 2011).

7.4.4 Joint Development Routines

Joint development routines involve practices related to joint product development and joint process improvement. Supplier integration levels are strengthened when buyers and suppliers work together in development tasks, improving collaboration between them and increasing resource base availability in innovation processes. When suppliers can participate directly in buyer's new product development processes, they can prepare suitable materials and resources to meet new requirements, identify design problems, and provide solutions, thus reducing the time and cost of new products and improving quality (Lau 2014). Figure 7.1 shows the conceptualization of supplier integration as a capability.

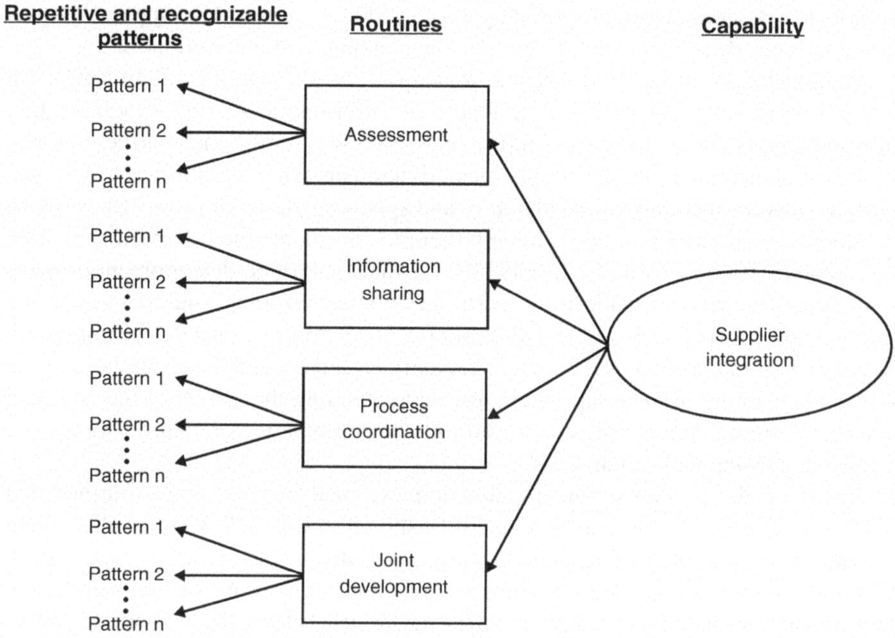

Fig. 7.1 Supplier integration capability

7.5 Supplier Integration as a Dynamic Capability

The concept of dynamic capabilities emerges from the concern that the resource-based approach (RBV) applies primarily to firms in stable environments, in which the resources and capabilities of the organization remain unchanged. However, when firms face dynamic environments, their competitive advantages quickly disappear, requiring the development of capabilities that let them to create, adapt, and reconfigure their resources to obtain a new competitive advantage (Ambrosini et al. 2009). A dynamic capability is defined as the ability of a company to integrate, build, and reconfigure internal and external competencies, in order to adapt to changes in the environment (Teece et al. 1997).

Organizational literature recognizes that the line separating dynamic capacities from operational capacities is inevitably blurred (Helfat and Winter 2011). Therefore, we consider that supplier integration capabilities can be operative or dynamic, depending on their intended use. To the best of our knowledge, Vanpoucke et al. (2014) were the first to conceptualize supplier integration as a dynamic capacity, suggesting that this type of integration serves a double purpose: (1) as an operational capability enabling communication and coordination between companies to streamline operations and delivery processes and (2) as a dynamic capability to support dynamic purposes, such as introducing new distribution channels. The authors use the framework proposed by Teece (2007) to assess that supplier integration combines three sub-capabilities; sensing, seizing, and transformation.

Vanpoucke et al. (2014) argued that in a buyer-supplier relationship, an integration-sensing capability is a result of information-sharing practices that inform partners about current and future physical flows, letting each partner to make better decisions that optimize supply chain dynamism. An integration-seizing capability results from collaborative planning and synchronization of processes between the company and its suppliers, allowing them to coordinate their activities. Finally, the capability to transform the supply chain is the result of co-development projects with suppliers, which enable joint learning, co-specialization, and changes in the internal processes of each partner, when supply chain processes need to be improved. Supply chain integration as a dynamic capacity let buyers and their suppliers to see what is happening in their supply chains, understanding the causes of the situation and transforming their supply chains through the combination of their resources and knowledge (Vanpoucke et al. 2014).

Based on the previous conceptualization, we analyzed the main routines that make up the supplier integration capability shown in Fig. 7.1. We classified them into the sub-capabilities of sensing, seizing, and transforming dynamic capabilities. Assessment routines involve activities such as the frequent use of the quantitative and qualitative metrics of supplier performance. Therefore, these routines enable buyers to know whether their requirements are met by their suppliers. Information-sharing routines involve patterns such as the exchange of information regarding plans and production capacities, inventory availability, and the forecast of demand. Therefore, both evaluation routines and information-sharing routines let firms to identify opportunities by sensing opportunities and taking advantage of them. We classified these routines as part of sensing capability, since sensing does not only involve investment in research activities, exploration of clients' needs, and technological possibilities but also it lets suppliers' needs to be understood (Teece 2007).

Process coordination routines involve activities such as joint planning, synchronization of activities, and structured problem-solving. We classified coordination routines as part of the sub-capability of seizing, since it allows organizations to apply external knowledge to their planning deployments, anticipating problems, and materializing opportunities (Kulp et al. 2004; Vanpoucke et al. 2014). Finally, joint development routines involve activities such as working together to improve inter-organizational processes, the involvement of suppliers in new product development processes, and supporting buyer activities with suppliers. We classified joint development routines within transformation capabilities because they enable buyers and their suppliers to learn from each other in reconfiguring their processes and implementing innovations that let them to create or adapt to changes in the environment. Figure 7.2 shows the conceptualization of supplier integration as a dynamic capacity.

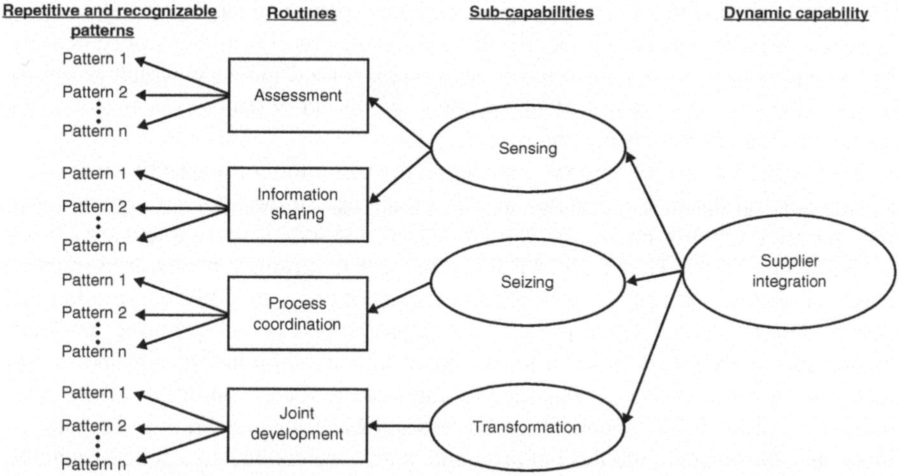

Fig. 7.2 Supplier integration as a dynamic capability

7.6 Research Methods

The connection between routines that make up supplier integration as a dynamic capacity and radical innovation has not been studied. Therefore, to analyze this link empirically, we used a qualitative methodology because it provides tools to study complex phenomena. Specifically, we used an exploratory case study in an organization that stands out in terms of both supplier integration and radical innovation. This organization is Herbes del Moli.

7.6.1 Research Context

The organic agro-food industry has experienced continuous and exponential growth worldwide. The most recent statistics suggest a market value of more than €60 billion with forecast growth expected to reach €100 billion in 2025. These figures highlight the importance of organic food production in the world. Quality and safety in this industry are continuously evolving. Quality is stimulated by product certifications, which ensure that ecological products are rigorously controlled. Therefore, the organic agro-food industry offers products with higher nutritional quality that lead to healthier foods and, at the same time, enormous respect for diversity and for the biological cycles of the soil (PRODESCON 2016; MAPAMA 2015).

The ecological paradigm has a global goal focused on achieving sustainability based on environmental, social, and economic dimensions, seeking maximum overlap (CEA 2015). According to Sanchez Pello (2014), consumers often show growing concern about environmental and health issues, which makes companies want to bring greater value to society by considering environmental sustainability, seeking an adequate balance between social and natural demands. Therefore, safety, healthy, and sustainability are essential in the organic agro-food industry, which promotes improvements at each stage of the product cycle and collaboration between the agents involved in the supply chain.

Herbes del Moli is a company founded at the beginning of the 1980s. It markets its products on the five continents and is a reference in the Spanish organic agro-food industry. Quality, environmental protection, and social commitment are Herbes del Moli's priorities. Herbes del Moli is a pioneer in aromatic plants used in products that meet ecological agro-industry requirements. In addition, it obtained FairWild certification, which proves that its production system promotes environmental sustainability and respect for the environment, and it invests a proportion of its profits in projects that are destined to improve the social conditions of the community. In order to understand how Herbes del Moli has managed to become an innovative company in the field of aromatic plants, we conducted a semi-structured interview with its technical director. Unless otherwise noted, quotations below come from this interview.

7.7 Findings

To perform our analysis, we focused on the four routines that make up supplier integration, as summarized in Fig. 7.1. In Herbes del Moli, there are information-sharing activities, especially focused on the sensing of changes in demand. "We often contact our suppliers to inform them if there are changes in demand, since we work with agricultural products, which do not let major changes to be made in production in the short term." Information is shared through meetings with suppliers and scheduled visits that enable new opportunities to be detected. "Information flows let us and suppliers to sense new opportunities."

It was also found that Herbes del Moli assesses its suppliers. These evaluations are mainly based on compliance with specifications, quality, and delivery. In addition, documentary audits are carried out to see whether the suppliers have the certifications that are required in the organic food industry. It is important to note that audits carried out in suppliers' facilities enable the identification of new opportunities for improvement: "from the findings found in the audits, we help our suppliers to implement improvements, providing them with any related information and knowledge that we have." According to the above, in Herbes del Moli information-sharing and assessment routines facilitate the development of products that require a high degree of new knowledge, since they let them sense changes in product demand, determine supplier needs, and assess whether they can meet the new requirements.

Herbes del Moli carries out coordination activities, such as joint planning with its main suppliers, which enables the company to materialize its innovation projects. "When a customer requires a new product, we contact suppliers whom we believe have the knowledge and ability to meet new requirements, and if we agree with one, we work together with the supplier and the customer, we plan production together, enabling the needs of the customer to be met, as well as the needs of the supplier so it can achieve the desired production capacity." Finally, in Herbes del Moli, suppliers have major responsibility for new product development, since they are the ones who materialize an important part of them. "We give them all the technical support we can. When it is a new product, neither suppliers nor ourselves have much experience, we bring together all the knowledge that we have gained from similar products, and we pool it with our suppliers to make sure production is successful." Joint development routines in Herbes del Moli are important to radical innovation, since through them, the firm improves its relationship with its suppliers, because suppliers feel that they are involved in the innovation process. Moreover, it lets them to take advantage of both firm and supplier knowledge, ensuring that both parties meet new requirements and obtain benefits.

Finally, it is important to note that when Herbes del Moli tries to pursue radical innovations, current suppliers are the first choice to be part of them. "Normally, we first contact our current suppliers, if none of them have the production capacity, then we look for new suppliers, but the first step is always with current suppliers." This shows that supplier integration represents an important dynamic capability that Herbes del Moli always takes into account when adapting, building, and reconfiguring its resources to implement radical innovation.

7.8 Discussion and Conclusion

Organizational theory suggests that radical innovation involves greater risks for the company. However, this degree of innovation is important because it enables companies to adapt to changes in the environment, ensuring their long-term survival. One of the key aspects that is highlighted for firms to survive in dynamic environments is the role of their suppliers in innovation processes, since if firms fail to ensure the sustainability of their supplies, they are doomed to failure in their adaptation attempts.

In this research, we have explored the connection between supplier integration and radical innovation. To do so, we used the framework of organizational routines and dynamic capabilities, examining how supplier integration is composed of simple routines that can build a dynamic capability. In this sense, we saw that the boundaries between capabilities and routines are blurred because capabilities are formed mainly by routines learned through training or experience, which confirms the perspective that routines stem from knowledge (Nelson and Winter 1982; Teece et al. 1997). The analysis of this connection was carried out in a Spanish organic agro-food firm, since organic firms need to develop radical innovations that let them to compete with conventional markets and follow an approach that ensures environmental, social, and ecological sustainability.

We found that supplier integration is mainly composed of four major routines: assessment, information-sharing, process coordination, and joint development. Additionally, through an analysis of the patterns that make up these routines, we determined how supplier integration as a dynamic capacity is formed. Based on the conceptualization made by Vanpoucke et al. (2014), we classified the identified routines, determining the micro-foundations of supplier integration as a dynamic capacity. Assessment and information-sharing routines are part of a package of routines that make up sensing capabilities. Process coordination routines are related to the seizing capability, and joint development routines are related to the transformation capability. Therefore, by performing these routines, firms can generate supplier integration as a dynamic capability.

Our analysis of Herbes del Moli reveals that this firm clearly performs the four routines identified above to achieve integration with its suppliers. Assessment and information-sharing routines enable the needs of suppliers to be sensed. Thus, they are able to help them cope with changes in the environment. Process coordination routines let joint planning between Herbes del Moli and its suppliers in such a way that any new opportunities sensed can be materialized. Finally, joint development routines improve the relationship between Herbes del Moli and their suppliers and facilitate the reconfiguration of resources by spreading knowledge and resources throughout the supply chain when radical innovations are carried out.

This research contributes to organizational theory, since it increases the understanding of supplier integration as a dynamic capacity through the study of micro-foundations from an organizational routines point of view. In addition, from our analysis, we discovered a deep-rooted connection between supplier integration and the development of radical innovation. The exploratory case study showed that radical innovation requires the four routines that make up supplier integration. Therefore, supplier integration as a dynamic capacity not only lets firms to adapt to environmental changes but also molds them through radical innovations carried out collaboratively between firms and their suppliers.

Additionally, this research shows that it is important to study supplier integration as a dynamic capacity from a bidirectional viewpoint, considering the ability of the company to assimilate external knowledge and adapt its resources but, at the same time, understanding how buyers provide knowledge and help their suppliers so they can adapt their resources to cater for new requirements. Exploring how buyers and suppliers achieve a balance between these activities would be a promising topic for future research.

It is also important to note that when radical innovations are undertaken, Herbes del Moli always attempts to work with current suppliers. However, when obstacles are found that impede collaboration with current suppliers, new suppliers that can meet the new requirements are sought. Future research could determine which main obstacles are preventing firms from carrying out radical innovations with their current suppliers. Moreover, it would be interesting to consider the customer integration role in the development of radical innovation, since this type of integration is recognized by improving a firm's ability to sense changes in the environment. Finally, future research could study the synergies between supply chain integration and quality management systems, especially in industries where the commitment to sustainability is high, since these industries have specific quality management systems that may require a greater level of integration in the supply chain.

Acknowledgments This research is part of the Project ECO2015-71380-R funded by the Spanish Ministry of Economy, Industry and Competitiveness and the State Research Agency. This is also co-financed by the European Regional Development Fund (ERDF).

References

Ambrosini, V., Bowman, C., & Collier, N. (2009). Dynamic capabilities: An exploration of how firms renew their resource base. *British Journal of Management, 20*(S1), S9–S24.

Anderson, M., & Katz, P. (1998). Strategic sourcing. *The International Journal of Logistics Management, 9*(1), 1–13.

Bonaccorsi, A., & Lipparini, A. (1994). Strategic partnerships in new product development: An Italian case study. *Journal of Product Innovation, 11*(2), 134–145.

CEA. (2015). Hacia una estrategia agroalimentaria sostenible para Vitoria-Gasteiz. Materiales basicos para un diagnistico participativo. *Centro de Estudios Ambientales, 1,* 1–183.

Chandy, R., & Tellis, G. (1998). Organizing for radical product innovation: The overlooked role of willingness to cannibalize. *Journal of Marketing Research, 35*(4), 474–487.

Chen, I. J., & Paulraj, A. (2004). Towards a theory of supply chain management: The constructs and measurements. *Journal of Operations Management, 22*(2), 119–150.

Day, M., Lichtenstein, S., & Samouel, P. (2015). Supply management capabilities, routine bundles and their impact on firm performance. *International Journal of Production Economics, 164,* 1–13.

Eisenhardt, K. M., & Tabrizi, B. N. (1995). Accelerating adaptive processes: Product innovation in the global computer industry. *Administrative Science Quarterly, 40*(1), 84–110.

Feldman, M. S., & Pentland, B. T. (2003). Reconceptualizing organizational routines as a source of flexibility and change. *Administrative Science Quarterly, 48*(1), 94–118.

Flynn, B. B., & Zhao, X. (2014). *Global supply chain quality management.* Boca Raton: CRC Press.

Flynn, B. B., Huo, B., & Zhao, X. (2010). The impact of supply chain integration on performance: A contingency and configuration approach. *Journal of Operations Management, 28*(1), 58–71.

Frohlich, M., & Westbrook, R. (2001). Arcs of integration: An international study of supply chain strategies. *Journal of Operations Management, 19,* 185–200.

He, Y., Keung, K., Sun, H., & Chen, Y. (2014). The impact of supplier integration on customer integration and new product performance : The mediating role of manufacturing flexibility under trust theory. *International Journal of Production Economics, 147,* 260–270.

Helfat, C., & Winter, S. (2011). Untangling dynamic and operational capabilities: Strategy for the (N) ever-changing world. *Strategic Management Journal, 3*(11), 1243–1250.

Huo, B., Zhao, X., & Lai, F. (2014). Supply chain quality integration: Antecedents and consequences. *IEEE Transactions on Engineering Management, 61*(1), 38–51.

Ittner, C., & Larcker, D. (1997). Product development cycle time and organizational performance. *Journal of Marketing Research, 34*(1) 13–23.

Kulp, S., Lee, H., & Ofek, E. (2004). Manufacturer benefits from information integration with retail customers. *Management Science, 50*(4), 431–444.

Lai, F., Zhang, M., Lee, D. ., & Zhao, X. (2012). The impact of supply chain integration on mass customization capability: An extended resource-based view. *IEEE Transactions on Engineering Management, 59*(3), 443–456.

Lambert, D., & Cooper, M. (1998). Supply chain management: Implementation issues and research opportunities. *The International Journal of Logistics Management, 9*(2), 1–20.
Lau, A. K. W. (2011). Supplier and customer involvement on new product performance. *Industrial Management & Data Systems, 111*(6), 910–942.
Lau, A. K. W. (2014). Influence of contigent factors on the perceived level of supplir integration: A contigency perpective. *Journal of Engineering and Technology Management, 33*, 210–242.
Leifer, R. (2000). *Radical innovation: How mature companies can outsmart upstarts*. Boston: Harvard Business Press.
Lin, R., Chen, R., & Chiu, K. (2010). Customer relationship management and innovation capability: An empirical study. *Industrial Management & Data System, 110*(1), 111–133.
Luo, X. (2002). Trust production and privacy concerns on the internet: A framework based on relationship marketing and social exchange theory. *Industrial Marketing Management, 31*(2), 111–118.
MAPAMA. (2015). Programa para la sostenibilidad integral de la industria agroalimentaria. *Ministerio de Agricultura, Alimentación Y Medio Ambiente, 1*, 1–35.
McDermott, C., & O'Connor, G. (2002). Managing radical innovation: An overview of emergent strategy issues. *Journal of Product Innovation Management, 19*(6), 424–438.
Mentzer, J., Min, S., & Zacharia, Z. (2000). The nature of interfirm partnering in supply chain management. *Journal of Retailing, 76*(4), 549–568.
Monczka, R. (2000). *New product development: Strategies for supplier integration*. Milwaukee: ASQ Quality Press.
Morgan, J., & Monczka, R. (1996). Supplier integration a new level of supply chain managment. *Purchasing, 120*(1), 110–113.
Müller, M., & Gaudig, S. (2011). An empirical investigation of antecedents to information exchange in supply chains. *International Journal of Production Research, 49*(6), 1531–1555.
Narasimhan, R., & Das, A. (1999). An empirical investigation of the contribution of strategic sourcing to manufacturing flexibilities and performance. *Decision Sciences, 30*(3), 683–718.
Narasimhan, R., & Kim, S. (2002). Effect of supply chain integration on the relationship between diversification and performance: Evidence from Japanese and Korean firms. *Journal of Operations Management, 20*(3), 303–323.
Näslund, D., & Hulthen, H. (2012). Supply chain management integration: A critical analysis. *Benchmarking: An International Journal, 19*(4/5), 481–501.
Nelson, R., & Winter, S. (1982). *An evolutionary theory of economic change*. Cambridge, MA: Harvard University Press.
Nyaga, G., Whipple, J., & Lynch, D. (2010). Examining supply chain relationships: Do buyer and supplier perspectives on collaborative relationships differ? *Journal of Operations Management, 28*(2), 101–114.
Peng, D. X., Schroeder, R. G., & Shah, R. (2008). Linking routines to operations capabilities: A new perspective. *Journal of Operations Management, 26*(6), 730–748.
Petersen, K. J., Handfield, R. B., & Ragatz, G. L. (2005). Supplier integration into new product development: Coordinating product, process and supply chain design. *Journal of Operations Management, 23*(3–4), 371–388.
PRODESCON, S. A. (2016). *Caracterización del sector de la Producción ecológica española En términos de valor y mercado, Referida al año 2015*. España.
Ragatz, G., & Handfield, R. (1997). Success factors for integrating suppliers into new product development. *Journal of Product Innovation Management, 14*(3), 190–202.
Ragatz, G., Handfield, R., & Petersen, K. (2002). Benefits associated with supplier integration into new product development under conditions of technology uncertainty. *Journal of Business Research, 55*(5), 389–400.
Rosenzweig, E. D., Roth, A. V., & Dean, J. W. (2003). The influence of an integration strategy on competitive capabilities and business performance: An exploratory study of consumer products manufacturers. *Journal of Operations Management, 21*(4), 437–456.

Sanchez Pello, P. (2014). *La industria alimentaria y el medio ambiente*. Retrieved from cincodias.elpais.com/cincodias/2014/06/13/empresas/1402687653_251085.html

Song, M., & Di Benedetto, C. A. (2008). Supplier's involvement and success of radical new product development in new ventures. *Journal of Operations Management, 26*(1), 1–22.

Soosay, C. A., Hyland, P. W., & Ferrer, M. (2008). Supply chain collaboration: Capabilities for continuous innovation. *Supply Chain Management: An International Journal, 13*(2), 160–169.

Stank, T., & Keller, S. (2001). Supply chain collaboration and logistical service performance. *Journal of Business Logistics, 22*(1), 29–48.

Swink, M. (1999). Threats to new product manufacturability and the effects of development team integration processes. *Journal of Operations Management, 17*(6), 691–709.

Swink, M., & Nair, A. (2007). Capturing the competitive advantages of AMT: Design–manufacturing integration as a complementary asset. *Journal of Operations Management, 25*(3), 736–754.

Swink, M., & Song, M. (2007). Effects of marketing-manufacturing integration on new product development time and competitive advantage. *Journal of Operations Management, 25*(1), 203–217.

Teece, D. (2007). Explicating dynamic capabilities: The nature and microfoundations of (sustainable) enterprise performance. *Strategic Management Journal, 28*(13), 1319–1350.

Teece, D., Pisano, G., & Shuen, A. (1997). Dynamic capabilities and strategic management. *Strategic Management Journal, 18*(7), 509–533.

Tempelaar, M. P., & Van Den Vrande, V. (2012). Dynamism, munificence, internal and external exploration-exploitation and their performance effects. In *Annual meetings of the Academy of Management*.

Thomas, D., & Griffin, P. (1996). Coordinated supply chain management. *European Journal of Operational Research, 94*(1), 1–15.

Vanpoucke, E., Vereecke, A., & Wetzels, M. (2014). Developing supplier integration capabilities for sustainable competitive advantage: A dynamic capabilities approach. *Journal of Operations Management, 32*(7), 446–461.

Walter, A., & Gemünden, H. G. (2000). Bridging the gap between suppliers and customers through relationship promoters: Theoretical considerations and empirical results. *Journal of Business & Industrial Marketing, 15*(2–3), 86–105.

Wang, Z., Huo, B., Qi, Y., & Zhao, X. (2016). A resource-based view on enablers of supplier integration: Evidence from China. *Industrial Management & Data Systems, 116*(3), 416–444.

Wong, C. Y., Boon-Itt, S., & Wong, C. W. Y. (2011). The contingency effects of environmental uncertainty on the relationship between supply chain integration and operational performance. *Journal of Operations Management, 29*(6), 604–615.

Wong, C. W. ., Wong, C. Y., & Boon-Itt, S. (2013). The combined effects of internal and external supply chain integration on product innovation. *International Journal of Production Economics, 146*(2), 566–574.

Zhao, X., Huo, B., Flynn, B., & Yeung, J. (2008). The impact of power and relationship commitment on the integration between manufacturers and customers in a supply chain. *Journal of Operations Management, 26*(3), 368–388.

Chapter 8
Personal Branding as a Knowledge Management Tool to Enhance Innovation and Sustainable Development in Organizations

Antonio Alonso-Gonzalez, Marta Peris-Ortiz, and Jose J. Cao-Alvira

8.1 Introduction

Nowadays the Personal Brand concept is a trending topic. Actors, singers, sportsmen, and models have become referents for important consumer groups, and as the big brands do, these characters perform actions to gain notoriety and increase the value of their Personal Brand. A Personal Brand allows an individual to become unique not only from the personal point of view, but also from the professional one, starting to manage a professional career as a company with business criteria rather than as a mere human resource, adding this way value to the personal and professional performance and results. Thus, Personal Branding is an evolution in conventional marketing that is called products and services applied positioning: an individual has a position in a determined environment, which is the way the professional is perceived by others, so positioning is an innate quality of humans. Therefore, Personal Branding consists of building trust and confidence within the individual and the environment, and the closer this relationship is, the stronger the bond will be.

This work seeks to analyze the possible uses of Personal Branding as a knowledge management tool within an organization in order to enhance innovation

---

A. Alonso-Gonzalez (✉)
EIAM-PRIME Business School, Universidad Sergio Arboleda, Bogotá, Colombia
e-mail: antonio.alonso@usa.edu.co

M. Peris-Ortiz
Department of Business Administration, Universitat Politècnica de Valencia, Valencia, Spain
e-mail: mperis@doe.upv.es

J. J. Cao-Alvira
City University of New York - Lehman College and EIAM-PRIME Business School, Universidad Sergio Arboleda, New York, USA
e-mail: jose.caoalvira@lehman.cuny.edu

M. Peris-Ortiz et al. (eds.), *Knowledge, Innovation and Sustainable Development in Organizations*, Innovation, Technology, and Knowledge Management,
https://doi.org/10.1007/978-3-319-74881-8_8

and sustainable development in our current society. Classic ideas of well-known authors defined the concept of economic and business development, like Schumpeter (1934), who argued that "economic development involves transferring capital from old businesses using established methods of production to businesses using new, innovative methods," or Penrose (2009) who thoroughly studied the development and growth of firms. It is important to highlight that from this point of view, other more innovative approaches could be taken into account such as Kovel (2007) and his vision of the end of capitalism in our current society. Therefore, this analysis can be complicated to some extent because Personal Branding is often interpreted as a personal tool, and its integration into organizations can generate resistance or rejection. Many Human Resources departments focus their policies on the classification of homogeneous attributes of human capital which is available in the organization. However, a Personal Brand is based on personal and professional differentiation and positioning toward the working environment, which can cause conflicts when this marketing philosophy is implemented within an organization.

This research paper aims to identify and relate to the aspects and elements that Personal Branding could encourage from innovation and a sustainable development perspective in an organization. In the theoretical background section, the Personal Brand concept will be discussed, as well as its impact on the human capital of an organization from the point of view of innovation and sustainable development. The methodology section will describe the procedures that were performed to analyze the impact, emphasizing in the results section the most important facts relating to Personal Branding and knowledge management, based on an intensive literature review and research. Finally, in the conclusions and future research section, the main results and findings made in the study will be summed up and discussed, giving some alignments for future research.

8.2 Theoretical Background

To succeed in this globalized and highly competitive world, professionals need to master the skills and competences related to a new discipline called Personal Branding. These competences will allow an individual to become a successful professional or entrepreneur, besides acknowledging his or her position, knowing how to relate to the society in which it operates with the working environment and how to promote their services as well (Valarezo-Paredes 2015). The analysis of the Personal Branding literature shows that there is a market need for this discipline which is strongly connected to an increase in job insecurity. The role of the marketer here is to identify new needs and develop ways to fulfill them. According to Zarkada (2011), "what is being turned into a commodity by the Personal Branding literature, however, is not really people but hope: the hope of standing out in the crowded spaces of urban modernity, the hope of being acknowledged, feeling unique and worthy of attention and most of all, the hope of finding meaning now that traditional values have been eroded and conspicuous consumption is fast losing ground as a panacea to obscurity and loneliness."

Trying to classify Personal Branding as a new discipline derived from conventional marketing, it can therefore be placed within the branding area. According to Perez-Ortega (2008), the concept of brand is defined by six dimensions: confidence, expectations, relationships, differentiation, value, and information.

- Information: Branding is a way of compressing information, condensing, and communicating the strength of a product, value, and position as concisely as possible – the combination of a name and a visual image, a personality that a consumer will recognize and interpret in the same way.
- Differentiation: The opposite of the notion of commodity is the concept of brand, a unique differentiation that makes consumers choose the product or service, positioning it in the consumer's mind as something differentiated with a unique identity.
- Expectations: A brand is a way of valuing what can be expected from something or someone. The brand is the sum of all the expectations and associations that are created in the mind of the audience, a promise that creates a lasting belief in the buyer. Branding is the way a company tells its customers what to expect.
- Value: The brand requires a product or service to be much more than raw material, labor, cost, or time. The brand increases the value because it provides security, prestige, and trust, which are intangibles that modify the perception of the customer to identify a brand as being of better quality or performance than another. Therefore, the goal of a brand is to leave a lasting impression relating to the nature of a company, a product, a service, or a process.
- Confidence: The brand is a definition of trust. Strong brands are based on consistency and clarity, so if a brand is well established and has been performing consistently for years, it generates confidence and it is able to withstand all kinds of challenges and possible mistakes.
- Relationships: For a brand, in order to grow and acquire a value, it must interact with its environment. A brand is the reflection of the relationship with its customers that a company maintains and expresses through its commitments, communicating and creating mutually lasting beneficial relationships.

According to Schawbel (2015), there are three types of brands:

- Corporate brands: The name, logo, and personality of a company. Corporate branding refers to using a corporate brand name to promote a product. For instance, Dove places its corporate brand name on bars of soap in the hope that a customer will purchase them based on their previous experience with the Dove brand, a commercial promoting it, or a friend who recommended it.
- Product/service brand: From a pen to a guitar to a car, a product is something a customer can touch. From lawn care to a haircut to legal advice, a service cannot be touched but nonetheless makes people's lives better. A corporate brand exists through products and services that carry its name. When consumers associate a particular product with a certain corporation, they tend to think of the corporation more positively.
- Personal Brand: Unlike a corporation, an individual does not need to put his or her name on a product to sell him or herself. A person is a type of product and, like a product, requires marketing. This means that that person can adopt several brand strategies in his or her own life, such as producing a website.

Thus, the brand concept not only applies to organizations or companies but it is also linked to places and people, naming this concept the Personal Brand and being studied by a marketing field called Personal Branding (Rico-Jerez 2013). Therefore, the individual communicates with his or her environment: the gestures, the look, the words, the clothes, the competences, the skills, and knowledge, this combination of elements produces a particular Personal Brand (Gonzalez-Pelaez 2014). According to Rampersad (2009), a Personal Brand influences positively on other's perceptions about values, qualities, singularities, and differences that a professional has, creating expectations in people's mind about what to expect if they work with the professional that holds the brand, building a memorable identity and maximizing the first considered option among the competition. Other studies such as the one conducted by Alonso-Gonzalez et al. (2017a) highlight the strong relationship between Personal Branding and leadership.

It is important to outline that just as a product or service brand can express a company brand, a Personal Brand can strengthen a company brand name. In the following sections, the most important concepts will be introduced in relation to Personal Branding and its implications and impact within an organization. In this literature review, we will describe various considerations that Personal Branding could have in relation to innovation and the sustainable development perspective as a tool of knowledge management.

8.2.1 The Personal Brand Concept

The Personal Brand idea has existed since the origins of mankind, but it has emerged as a concept in the last decade as an evolution of the corporate branding term applied to individuals (Ramos-Redondo 2015). Thus, from a historical point of view, the term or concept known as Personal Branding is completely new. The origins of this term are located in the individualistic society of the 1980s, and it was catalyzed by the emergence of information technology in the 1990s, which allowed marketing to be customized. Therefore, new market niches were sought to help differentiate customers even more and coupled with the historical era of globalization-capitalism, which helped the development of new forms of consumption. Based on this trend, the initial concept of Personal Branding was raised, considering that each individual has the opportunity to create his or her own Personal Brand and to maximize the client's satisfaction, personal and professional development, integrity, and personal growth, in order to fulfill goals and objectives as a lifetime project (Torres-Leaman 2011).

It was Peters (1997) who started the concept of Personal Branding through his article published in 1997 in Fast Company magazine called *The Brand Called You*, introducing the concept of what is necessary to be the CEO of "Me Inc.," encouraging readers to think about what differentiates them from the rest of their colleagues and competitors. According to this author, the professional is the owner of his own company: "Regardless of age, regardless of position, regardless of the business we happen to be in, all of us need to understand the importance of branding.

We are CEOs of our own companies: Me Inc. To be in business today, our most important job is to be head marketer for *The Brand Called You*. Other authors that can be cited from the last century were Armstrong and Wai Yu (1997), authors of *The Persona Principle: How to Succeed in Business with Image Marketing*, where the business case study method is used to compare corporate brands with the incipient phenomenon of Personal Branding. Again, Peters (1999) wrote in his book *The Brand You 50 (Reinventing Work): Fifty Ways to Transform Yourself from an Employee into a Brand That Shouts Distinction, Commitment, and Passion!* and continues to enhance the Personal Brand concept and prepare it for the next century, consequently becoming a leader in this field of study.

In this century, Montoya and Vandehey (2002) in his book *The Personal Branding Phenomenon: Realize Greater Influence, Explosive Income Growth and Rapid Career Advancement by Applying the Branding Techniques of Michael, Martha, and Oprah* expands on the issue of Personal Branding as a strategy to improve the perception that people have about the individual. Subsequently, it was Drucker (2005) who actively encouraged the growth of the Personal Branding movement as a new concept with his article *Managing Oneself*. In his contribution, the author emphasized the importance of identifying the strengths of an individual and focusing on consistently improving them, because as he says "it takes more energy to improve from incompetence to mediocrity than to improve from first-rate performance to excellence." For the author it is also important to figure out how the individual can perform, therefore not everybody learns in the same way, and it is necessary to understand this not only identifying outstanding strengths but also realizing how these strengths can be improved. As Drucker says "Do not try to change yourself, because you are unlikely to succeed. Work to improve the way you perform."

In recent years other authors who contributed to the Personal Branding concept were Foglio (2007), who defined in his book *Il Marketing Personale. Il self-marketing della persona e della carrier*, the concept of self-marketing, focusing on its applications to develop a successful professional career. Perez-Ortega (2008) in his book *Personal Brand: Becoming the Preferred Option* discusses the current labor market situation and the importance for a professional of having a strong Personal Brand. The author builds, explains, and recommends his Personal Branding model called the DNA model, composed of four outside elements, six interior elements, and six description elements. Russo and Bustreo (2015) published the book *Self Marketing per le professioni. Un percorso formativo per i talenti dell'alta formazione*, in which the authors explain the importance of self-marketing in high-profile career development. As recent contributions, Schawbel (2015) in his book published *Me 2.0: 4 Steps to Building Your Future* explains the strategies to help a professional become a successful individual in today's world, providing tools to define and achieve an individual's goals. The author describes the history of Personal Branding, explaining "how the Internet revolutionized career development and how individuals can leverage social media for personal empowerment, self-management, and networking." The author includes a four-step process for discovering, creating, communicating, and maintaining a Personal Brand and also taking into account the element of time.

To conclude this literature review about the evolution of the Personal Branding concept, it is cited the definition of Rampersad (2009) in reference to how a Personal Brand should be: authentic, reflecting a true personality and being built from dreams, vital assets, values, singularities, gifts, passions, specializations, and things that individuals like to do. According to the author, if a professional creates a Personal Brand in this organic, authentic, and holistic way, it will be distinctive, relevant, meaningful, inspiring, convincing, lasting, clear, persuasive, and memorable. The individual will develop a satisfying life, attracting others who are aligned and increasing the ability to deliver maximum performance.

8.2.2 Personal Branding in an Organizational Context

In an economy where the standardization of processes is constant, with the corresponding homogenization of products and where workers are treated as commodities or white-label professionals, it is essential to differentiate and to recognize the added value that a professional contributes to his company, environment, or market. The best way to do this is to discover and communicate the differences that perceive them as outstanding or excellent professionals, and this should be the basic strategy of Personal Branding. As professionals, it is necessary to think about the personal job as a product or service, about the profession as a company, and about the identity as a brand (Perez-Ortega 2008). Price will cease to be a factor for the customer to make a decision, buy a product, or hire a service, and therefore a Personal Brand is required in order to communicate all the elements that make up a professional in the market who is distinct from the competition (Estalella 2011).

Relating the concept of Personal Branding within the organizational context and environment, Orjuela-Cordoba (2008) explains that the most important asset for a person is the name, because every individual is a brand. For this reason, the author recommends performing marketing techniques, for example, market intelligence, sales management, and branding, in order to improve the performance at a professional level. These initiatives are not only effective for freelance or businesspeople but also for company employees. Thus, it is important for the author to apply Personal Branding inside the organizations to maximize the effect on performance and results. According to Sumarokova and Zaharenko (2012), when a professional is forming the basis for his career, he or she appears simultaneously as a creator and as a salesman of his or her own workforce. Therefore, it is necessary to determine correctly what needs are required within the organization that will be in great demand or which problems could be successfully resolved by the professional. Even having valuable skills and competences, it is possible to fail if the professional does not know how to market or communicate them to their managers or colleagues. Therefore, it is necessary in a person's career to be a master of Personal Branding methods. Personal Branding tools will help a person to achieve their career goals in an organization and increase their value in the labor market.

Companies require more and more competitive individuals for their staff and in this environment professionals must learn to market themselves as a brand in order to gain visibility within the company, demonstrating all the advantages and value that makes them attractive and desirable to the company (Infante-Luey 2010). Business leaders must be distinguished by their excellence, which will only be achieved with a strong Personal Brand based on technical skills, a humanistic background, leadership skills, and solid human values. A combination of a holistic vision and previous experience in lower positions is recommended, in order to combine memories from the past with a present vision of the big picture. Other skills such as self-confidence, communication skills, and a capacity to generate trust and empathy are also desirable (Rico-Jerez 2013). Personal Branding requires professionals to behave as companies within the system and organizations, such as entrepreneurs and intra-entrepreneurs. This must not be considered a question of confrontation but of collaborating as equals, and consequently the so-called psychological contract ceases to exist. Therefore, Personal Branding can change the conventional rules and turn the center of gravity from the organizations to the professionals and the people (Perez-Ortega 2008).

8.2.3 Personal Branding as an Innovation and Sustainable Development Tool in Organizations

According to Sumarokova and Zaharenko (2012), "a great number of modern companies came to understanding that their competitive advantages depend on human and social capital. In the information society and highly technological economy of the XXI century, some elements like rates of development, quality of goods and services are increasingly predetermined by people, their social and human capital. Therefore, central objectives and priorities of the current contemporary social and economic development are the creation of conditions which may facilitate development and increase of human and social capital. From the very beginning, the development of these concepts has been based on integration between traditional economic concepts and sociological and political science." This line of thought is convergent with the study performed by Cheng and Huizingh (2014) which links the enhancements of innovation environments in a company with the role of strategic orientation. Further works developed by Alonso-Gonzalez et al. (2017a, 2017b) focus on innovation from the perspective of higher education institutions and their role in the generation of initiatives to enhance creativity, innovation, and sustainable development scenarios.

Thus, the approach of building a Personal Brand requires a change of focus, where a professional must become the true owner of their work, considering the company that pays the payroll as a "star client." However, to discover and build the Personal Brand, it must be identified who the professional is, what the professional knows, and what the professional wants to do, in order to progress in their professional career and improve skills and knowledge. The success of professionals is no longer guaranteed by their CV. The value of people and the differential of development are a consequence of their Personal Brand, which will be identified by their soft skills, for example, their ability to relate, to transmit innovation, and to promote personal uniqueness among their colleagues, managers, clients, suppliers, partners, friends, and family (Perez-Ortega 2008). Consequently, companies must consider this new discipline of Personal Branding as part of their knowledge management policies within the Human Resources department, in order to enhance personal proactive innovation, dynamic capabilities, and sustainable development for their employees. For these Human Resources departments, it is important to start by identifying the qualities or characteristics that make a professional distinctive from competitors or even colleagues (Peters 1997).

According to Drucker (2005), it is also important that values professed by a professional must be equal, similar, or at least convergent with the ones shared by the company, clients, or environment. The author affirms that other important questions that companies should answer regarding their employees is the type of environment in which their professionals perform better (a stressed and changing environment or a highly structured and predicable one) and the kind of structure to work in (a big company or a small one), because only a small number of people work well in all kinds of environments. Drucker explains that "there are professionals who are very successful in large organizations but flounder miserably when they are moved into smaller ones, the reverse being equally true." Therefore, to locate the right professionals at the right places within an organization is mandatory in order to create a proper environment which enhances innovation, creativity, dynamic capabilities, and sustainable development.

However, as Baruk (2012) highlights in her study, it is also important to take into account that the external image of an employer is shaped based not only on an employer's activities in relation to prospective employees but also on Human Resources activities targeted at current employees, who transmit their opinions to the participants of the external labor market via informal communication channels. Therefore, the image of an organization as an employer is a result of formal activities undertaken by an organization and informal messages transmitted by current employees to other people. That is the reason why it is so important for an organization to eliminate irregularities in its Human Resources policies, in order to win the crucial backing of its employees and to rely on them to transmit positive communications about its operations as an employer. In this way employees turn into ambassadors not only of the employer image but also of the overall image of the organization, innovation processes, sustainable development initiatives, shared values, and culture.

Other important issues that organizations should consider in relation to the application of Personal Branding as a knowledge management tool is the use of information and communication technologies – ICTs. According to Cortes-Arevalo et al. (2016), in the current globalized world, the gap between physical and virtual communications has been narrowing, facilitating interpersonal relations in distance and other phenomena derived from technological advances and competitiveness, because it is more important to stand out among others. In this line of thought, Schawbel (2015) argues that the labor market has changed considerably, "thanks to the Internet, traditional methods of communication and career development are evolving. As a result, businesses of all sizes cannot survive by legacy systems alone and will not adapt to these changes without harnessing the power of online technology. In addition, the Internet itself is changing. It is no longer just a place to research and discover information, to purchase products and send emails. Nowadays it is a tool to connect with people from all over the world, regardless of age, race, gender, religion, and social class, to meet, discuss ideas, and do business. Thus, a new breed of worker is emerging, who feels energized and empowered by this new world that the Internet has created and is eager to make a difference." Therefore, for current companies it is important to be able to understand and manage these new scenarios and to make the most of modern technology, including creating powerful Personal Brands for their employees and helping them stand out and achieve their goals.

8.3 Methodology

The methodology applied in the present study to analyze Personal Branding impact in an organization's innovation and sustainable development as a knowledge management tool was based on an extensive analysis of the literature in order to identify some important aspects that companies' need to take into account. This analysis was conducted in order to set a relation of best practices from the Personal Branding perspective that could be implemented by an organizations Human Resources department in order to enhance these factors.

As it was established in the theoretical background section, three different stages were developed to conduct the literature review in order to perform an approach from the general term of Personal Branding, concerning its impact on an organization. Therefore, it is important to discuss the effects that this new marketing discipline could achieve regarding innovation and sustainable development:

- The first part of the literature review was an analysis of the Personal Branding evolution concept, taking into account the studies performed by Drucker (2005), Armstrong and Wai Yu (1997), Peters (1997), Peters (1999), Montoya and Vandehey (2002), Foglio (2007), Andres Perez-Ortega (2008), Rampersad (2009), Torres-Leaman (2011), Ramos-Redondo (2015), Russo and Bustreo (2015), and Schawbel (2015).
- The second part of the literature review allows us to understand the beneficial impact that Personal Branding could achieve in general terms for an organization, taking into account the studies performed by Orjuela-Cordoba (2008), Perez-Ortega (2008), Infante-Luey (2010), Estalella (2011), Sumarokova and Zaharenko (2012), and Rico-Jerez (2013).
- The third part focuses on Personal Branding from the knowledge management dimension within an organization and its effects in terms of innovation, dynamic

capabilities, and sustainable development, taking into account the studies performed by Peters (1997), Drucker (2005), Perez-Ortega (2008), Baruk (2012), Sumarokova and Zaharenko (2012), Cheng and Huizingh (2014), Schawbel (2015), Cortes-Arevalo, Sanchez-Saenz and Alonso-Gonzalez (2016), and Alonso-Gonzalez et al. (2017a, 2017b).

Other authors who complemented this analysis of the literature were Zarkada (2011), Rico-Jerez (2013), Gonzalez-Pelaez (2014), Valarezo-Paredes (2015), and Alonso-Gonzalez et al. (2017a). It is important to note that the sources found decreased as the analysis was conducted. A substantial amount of material was found in the case analysis of the Personal Branding evolution concept. More limited material was found when the impact of this discipline was studied in the organizations and little if any when analyzing the impact of Personal Branding in the processes of innovation, dynamic capabilities, and sustainable development of organizations.

8.4 Results

As it has been explained in the literature review, Personal Branding can be defined as the process of creating the Personal Brand, a method that will identify the skills and strengths that define the person and make them stand out in comparison to other professionals. A Personal Brand serves to communicate the personality and professionalism of an individual in an environment, in order to facilitate the identification of his or her skills by others, knowledge, and experience and thus choose him or her as an ideal candidate. Therefore, proper Personal Branding could be defined as a tool that disseminates and promotes the brand through online and offline channels that are within reach; it communicates why it stands out and why it is different, in a homogeneous, competitive, and changing environment.

8.4.1 Personal Branding Results in Individuals

A Personal Brand is not only an element to succeed in a professional career but also in personal life, due to the exercise of introspection required to identify attributes, strengths, and weaknesses and how to deliver them to the people who belong to the individual's environment. Personal Branding helps to achieve two important aspects for a person's life: happiness and success. Happiness could be defined as a perception of individual welfare, and success could be described as acts of outstanding performance and recognition. Therefore, Personal Branding is a tool to achieve a successful career development from the personal and professional perspective. Some of the effects identified are the following:

- Personal Branding structures a better life's project, growing in a personal and professional way. It can help to clarify idea and, set goals and priorities, in order to be more efficient, and in the process have a more complete, useful, and productive life for oneself and others. When Personal Branding is implemented by an individual, it helps to set career objectives and goals, acquiring a better perception of oneself and professional objectives and therefore improving time management efficiency.
- It is important to highlight that a Personal Brand does not guarantee that all the proposed objectives can be achieved, but it considerably increases the chances of success in new projects.
- Personal Branding increases leadership soft skills such as authority, influence, respect, appreciation, reputation, prestige, and decision-making. It also develops other soft skills such as confidence, conscience, security, determination, creativity, initiative, and flexibility.
- Personal Branding can help professionals to measure and optimize their output value and selling opportunities. A strong Personal Brand helps to communicate the professional's value proposal, through a clear and defined message to the target audiences.
- Personal Branding enables a professional to be considered as the first buying option in the market, offering something relevant, different, and visible and getting recognition, respect, and reputation for the professional's talents. It improves the personal image and attractiveness for the working environment.
- Personal Branding helps a professional know how to better compete in the market and how to differentiate from the competitors, thus offering greater value.
- Personal Branding sets preferences to search for the right customers and contacts.
- Personal Branding reduces uncertainty and offers better preparation to tackle dynamic changes and obstacles, thus becoming individually speaking more adaptive, innovative, and willing to learn and improve.

8.4.2 Personal Branding Results in Organizations

The factors that influence the management of the Personal Brand are the same that apply for the branding of a company. Within each brand, different goals are marked and explored, as well as mission, vision, or values. With these items, each brand is consolidated according to its interests and qualities, since the qualities of each brand also retain some necessary differentiation. Therefore, Personal Branding can be a feasible tool to improve a companies' and an employee's performance, as long as it is supported and developed in the proper way. The following are the anticipated beneficial results for a company that enhances Personal Branding for its employees:

- Personal Branding helps to set corporate goals to accomplish rather than job positions, gives support on developing personal objectives in the organization rather than career plans, and shares values rather than a corporate culture.
- Personal Branding will help an employee in his or her career promotion within the organization and increase his or her professional value in the labor market.
- Personal Branding encourages credibility, confidence, and perception of work quality and projects, maximizing work satisfaction in a company and the organization, which makes it more attractive to the labor market, increasing the chances of better salaries and decreasing employees' turnover rates.
- Personal Branding increases teamwork (as employees know better what the others can offer), efficiency and work output from partners and colleagues (changing their minds from a commodity-type professional to a free-agent world).
- Personal Branding helps to discover, communicate, and position the competitive advantages that make employees perceived as outstanding or excellent professionals in their positions within their companies and Human Resources departments. As professionals, it is necessary to think about the personal job as a product or service delivered to the company, about the profession as a microenterprise associated with the organization, and about the employee's identity as a brand or product in partnership with the employer.
- Personal Branding allows a professional inside an organization to detect and determine properly the needs that will be in great demand or the problems that could be successfully solved by the professional. This posture maximizes performance levels in processes where employees with a strong Personal Brand are involved.

8.4.3 Personal Branding Results as an Innovation and Sustainable Development Tool

The Personal Branding discipline is not only a matter of improving image, appearance, voice, tone, modulation, gestures, or clothing. It is a matter of aligning personal knowledge, skills, and values of the professionals with the processes, values, and culture of the organization, in addition to the improvement in initiative, motivation, and creativity. Therefore, Personal Branding should be considered as an innovation and sustainable development tool within organizations, through a proper knowledge management policy directed by the companies' Human Resources management. The desirable effects of this Personal Branding discipline in relation to innovation and sustainable development are listed below:

- Personal Branding could be seen as a knowledge management tool to help an organization to achieve competitive advantages, which are needed to succeed in our current information society and highly technological economy of the twenty-first century. As most of these competitive advantages depend on human and social capital, as well as their innovation and sustainable development processes, Personal Branding could enhance these needs and help to accomplish companies' strategic goals and objectives.

- Professionals with a strong Personal Brand awareness start to work day by day to improve as individuals but always strive to meet the needs in order to satisfy the actors of the professional and personal working environment. This implies proactive dynamic capabilities, innovation, and sustainable development tasks which are beneficial for the company as a whole.
- A well-developed Personal Brand for a professional within an organization will improve the performance of professionals through the change of the mentality from the employee idea to the microenterprise attitude. Therefore, a Human Resources department that is concerned about the importance of Personal Branding and enhances its development as a knowledge management tool can obtain higher performance levels, dynamic capabilities, innovation rates, and sustainable development results from a favorable Personal Branding environment.
- A company that enhances Personal Branding as a knowledge management tool from its Human Resources area can enhance synergies and cooperation with colleagues within an organization in specific tasks related to processes optimization, new ideas generation, problem-solving, creativity, innovation, dynamic capabilities, and sustainable development.
- The increase in commitment to the organization and the establishment of a strong internal customer loyalty to the values and culture of the company that Personal Branding can encourage could result in an employees' higher level of satisfaction in their working environment. This can achieve a proactive innovative culture for the professionals who always look for optimization in processes, creative solutions, and a well-balanced development with an ongoing vision of improvement.
- Personal Branding must be a knowledge-based tool to optimize the location of the right professionals within an organization, in order to create the right environment which enhances innovation, creativity, and sustainable development.
- Positive and constructive communications in internal and external channels related to the organization's operations, processes, and projects are strongly recommended. From the Personal Branding perspective, employees turn into the ambassadors not only of their own image but also of the employer image and the overall image of the organization, innovation processes, sustainable development initiatives, dynamic capabilities, shared values, and culture.
- This enhancement process of innovation, creativity, and sustainable development through Personal Branding that is encouraged by the Human Resources department as a knowledge management tool must be strongly supported by information and communication technologies – ICTs. These new technologies must act as a catalyst for all the processes, maximizing dynamic capabilities and the beneficial effects that Personal Branding can achieve for an organization.

8.5 Conclusions and Future Research

As it has been explained in the literature review, Personal Branding can be defined as the process of creating the Personal Brand, that is, a method that will identify the skills and strengths that define the person and make them stand out in contrast to other professionals. A Personal Brand serves to communicate the personality and professionalism of an individual in an environment, in order to facilitate the identification of their skills, knowledge, and experience and select them as ideal candidates. Therefore, a proper Personal Brand could be defined as a tool that disseminates and promotes the brand through online and offline channels that are within reach and communicates what stands out and is different, in a homogeneous, competitive, and changing environment.

The methodology applied in the present study to analyze the Personal Branding impact of an organization's innovation and sustainable development as a knowledge management tool was based on an extensive analysis of the literature in order to identify some important aspects for a company to take into account. This analysis was conducted in order to set a relation of best practices from the Personal Branding perspective that could be implemented by an organizations Human Resources department to enhance these factors. To perform this, three different stages were developed to conduct the literature review in order to perform an approach from the general term of Personal Branding to its impact on an organization and describe the effects that this new marketing discipline could achieve in relation to innovation, dynamic capabilities, and sustainable development.

The first part of the literature review was an analysis of the Personal Branding concept evolution, identifying some interesting results from the individual point of view. When Personal Branding is implemented by an individual, it helps to set career objectives and goals, get a better perception of oneself and the objectives as professional, and improve time management efficiency. However, it is important to highlight that a Personal Brand does not ensure that all the objectives proposed can be achieved, but it considerably increases the chances of success in new projects. Other important results are that Personal Branding increases leadership soft skills such as authority, influence, respect, appreciation, reputation, and prestige. Decision-making is also encouraged and other soft skills such as confidence, conscience, security, determination, creativity, initiative, and flexibility. From the value proposal point of view, Personal Branding can help professionals to measure and optimize their output value and selling opportunities, communicating the professional's value proposal through a clear and defined message to the target audiences. This will enable a professional to be considered as the first buying option for their market through recognition, respect, and reputation. Personal Branding helps to differentiate from the competitors, offering greater value. It also helps in setting preferences to search for the right customers and the right contacts, reducing uncertainty, and offering a better preparation to tackle dynamic changes and obstacles.

The second part of the literature review allows us to understand the beneficial impact that Personal Branding could achieve in an organization, highlighting its implication to change the mentality of how employees deal with their work, tasks, and responsibilities, as well as career promotion opportunities and their value in the labor market, changing their minds from a commodity-type professional to a free-agent world. Personal Branding also helps to discover, communicate, and position the competitive advantages that distinguish employees as outstanding or excellent professionals in their positions within their companies and Human Resources departments. It can also encourage credibility, confidence, and perception of work quality and projects, maximizing work satisfaction and reducing turnover rates. Higher levels of teamwork, synergies, efficiency, and performance are also achieved, allowing employees to properly detect and determine within the organization the needs that will be in a great demand or the problems that could be successfully solved by the professional.

The third part focuses on Personal Branding from the knowledge management dimension within an organization and its effects in terms of innovation, dynamic capabilities, and sustainable development. In this section Personal Branding was identified as a tool that can help a company to achieve competitive advantages needed to succeed in the current highly competitive world, due to the strong dependence of these competitive advantages for the human and social capital, encouraging proactive innovation, dynamic capabilities, and sustainable development tasks which are beneficial for the company as a whole. Then, a Human Resources department that is concerned about the importance of Personal Branding and enhances its development as a knowledge management tool can obtain higher performance levels, better optimization in different processes and generation of new ideas, and better problem solving, creating and innovating an indoors sustainable development that result in a favorable Personal Branding environment. The Human Resources department can achieve these results by using Personal Branding as a knowledge management tool to optimize the location of the best professionals in the right places within an organization, increasing commitment to the organization and establishing a strong internal customer loyalty to the values and culture of the company and raising employees' satisfaction in their working environment. It is important to highlight that in order to achieve this innovation, creativity, and sustainable development results, Personal Branding must be strongly supported by positive and constructive communications in both internal and external channels, as well as suitable information and communication technologies – ICTs infrastructure, which act as a catalyst for all the processes, maximizing the beneficial effects that Personal Branding can achieve within an organization.

In conclusion, this study highlights that as far as Personal Branding is concerned, structural and departmental changes should occur in organizations. The Human Resources department should change its name to human talent, and instead of being a selector, coordinator, and supervisor of the work and administrative procedures of employees, it should become an agent to enhance the knowledge, dynamic capabilities, and skills of the company's human capital. There are many reasons and advantages to build and develop a strong Personal Brand, from the personal to the professional point of view.

As limitations of the present study, the absence of any previous studies which relate the Personal Branding concept with innovation, dynamic capabilities, or a sustainable development perspective within the knowledge management dimension must be pointed out. Therefore, this study could set a starting point from which other studies can explore deeper the effects that this discipline could contribute to the use of best practices in an organization. As future lines of research, the authors of the present paper will continue to explore the application of Personal Branding strategies and techniques within an organization, to analyze the different impacts and effects that the use of Personal Branding can achieve not only in innovation, dynamic capabilities, and sustainable development but also expand to other aspects such as employees performance, organizational culture, working environment, and staff turnover.

References

Alonso-Gonzalez, A., Alamo-Hernandez, P., & Peris-Ortiz, M. (2017a). Guardiola, Mourinho and Del Bosque: Three different leadership and personal branding styles. In M. Peris-Ortiz, J. Alvarez-Garcia, & M. C. Del Rio-Rama (Eds.), *Sports management as an emerging economic activity* (pp. 329–344). Springer International Publishing ISBN: 978-3-319-63907-9.

Alonso-Gonzalez, A., Plata-Rugeles, D., Peris-Ortiz, M., & Rueda-Armengot, C. (2017b). Entrepreneurial initiatives in Colombian Universities: The innovation, entrepreneurship and business Center of Sergio Arboleda University. In M. Peris-Ortiz, J. A. Gómez, J. M. Merigó-Lindahl, & C. Rueda-Armengot (Eds.), *Entrepreneurial Universities* (pp. 151–163). Springer International Publishing ISBN: 978-3-319-47949-1.

Armstrong, D., & Wai Yu, K. (1997). *The persona principle: How to succeed in business with image marketing*. Simon and Schuster, New York, USA.

Baruk, A. I. (2012). The role of personal marketing in forming the image of a science organization. *Prace Instytutu Lotnictwa, 226*(5), 71–88.

Cheng, C. C., & Huizingh, E. K. (2014). When is open innovation beneficial? The role of strategic orientation. *Journal of Product Innovation Management, 31*(6), 1235–1253.

Cortes-Arevalo, C., Sanchez-Saenz, S., & Alonso-Gonzalez, A. (2016). La marca personal como elemento decisorio de un buen maestro. *Revista Libre Empresa, 13*(2), 73–83.

Drucker, P. F. (2005). Managing oneself. *Harvard Business Review, 83*(1), 100–109.

Estalella, J. (2011). Claves para construir y comunicar una marca personal. *Iuris, 11*(03), 31–35.

Foglio, A. (2007). *Il marketing personale. Il self-marketing della persona e della carriera* (Vol. 665). FrancoAngeli, Milan, Italy.

Gonzalez-Pelaez, M. (2014). La Comunicación, la Marca Personal, la Imagen y el Lujo a través de la Historia. Desde la Roma Clásica hasta nuestros días de Innovación. In Mendieta-Ramirez, A. (Ed.), *Tendencias de Vanguardia en Comunicacion* (pp. 211–229) Spain: Vision Libros, Madrid.

Infante-Luey, I. A. (2010). *La Marca de servicio personal como factor de éxito en las fusiones o adquisiciones. Caso: fusión de Bellsouth y Telefónica Móviles*. Master's thesis, Lima: Universidad Nacional Mayor de San Marcos.

Kovel, J. (2007). *The enemy of nature: The end of capitalism or the end of the world?* London: Zed Books.

Montoya, P., & Vandehey, T. (2002). The personal branding phenomenon: Realize greater influence, explosive income growth and rapid career advancement by applying the branding techniques of Michael. In *Martha & Oprah*. USA: Peter Montoya Inc.

Orjuela-Cordoba, S. (2008). Marketing Personal: La marca soy yo. *DIRCOM, 2008*(76), 21–24.
Penrose, E. T. (2009). *The theory of the growth of the firm*. New York: Oxford University Press.
Perez-Ortega, A. (2008). *Marca Personal: Cómo convertirse en la opción preferente*. Madrid: ESIC Editorial.
Peters, T. (1997). The brand called you. *Fast company, 10*(10), 83–90.
Peters, T. (1999). The Brand You 50 (Reinventing Work): *Fifty ways to transform yourself from an employee into a brand that shouts distinction, commitment, and passion!*. New York: Knopf.
Ramos-Redondo, A. A. (2015). *Marca personal y redes sociales: el modelo del éxito*. Bachelor's thesis, Madrid: Universidad Pontificia Comillas.
Rampersad, H. (2009). *Tu marca personal* (Vol. 1). Madrid: LID Editorial Empresarial.
Rico-Jerez, M. (2013). La marca personal como seña de identidad del directivo: competencias distintivas. *Revista de Dirección y Administración de Empresas., 2013*(20), 57–73.
Russo, V., & Bustreo, M. (2015). *Self marketing per le professioni. Un percorso formativo per i talenti dell'alta formazione*. Milano: FrancoAngeli.
Schawbel, D. (2015). *Me 2.0: 4 steps to building your future*. New York: Diversion Books.
Schumpeter, J. (1934). *The theory of economic development*. Cambridge, MA: Harvard University Press.
Sumarokova, E. V., & Zaharenko, I. K. (2012) *Self marketing as the basis of human and social capital forming in a modern company*. Working Paper.
Torres-Leaman, L. G. (2011). *Propuesta de un plan de marketing para una marca personal*. Master's dissertation, Buenos Aires: Universidad Nacional de la Plata.
Valarezo-Paredes, E. F. (2015). *Desarrollo de competencias de marketing personal como ventaja competitiva para los futuros profesionales ecuatorianos* (p. 209). Observatorio de la Economía Latinoamericana.
Zarkada, A. (2011). The personal branding phenomenon: Pushing epistemological boundaries or desperately marketing marketing? *Proceedings 16th International Conference on Corporate and Marketing Communications: The New Knowledge Globalization era Future Trends Changing Corporate and Marketing Communications* (pp. 146–156). Athens University of Economics and Business.

Chapter 9
Expatriation Knowledge Management: The Role of Openness to Change and Work Engagement

Silveli Cristo-Andrade, Bruno Felix, and Emerson Wagner Mainardes

9.1 Introduction

In an increasingly global market, the internationalization of companies has been one of the strategies adopted to grow (Engle et al. 2015; Vianna and Souza 2009). Due to this strategy being recurring, either with the opening of an overseas headquarters, an entry into a joint venture, or an acquisition or merger, any of these decisions may lead a large number of professionals to become expatriates, i.e., an employee going to work in a foreign country (Baruch and Altman 2002; Black et al. 1991; Pereira et al. 2005; Tharenou and Caulfield 2010).

Despite being singled out as one of the main competitive advantages of today (BGRS 2016), it is precisely this act of sending the professional abroad that reveals a concern of the organizations, the high cost of sending, maintaining, training, and repatriating an expatriate (Caligiuri and Lazarova 2001; Kulkarni et al. 2010; Vidal et al. 2008). It is estimated that the cost of a single expatriate mission can reach up to 1.2 million dollars (Loes 2015). Keeping this in mind, the importance of the assertiveness of professional selection for this type of international task is evident.

Therefore, for organizations to optimize such high investments, it is necessary to search for professionals who succeed in their missions. Loes (2015) supports that the incorrect recruitment of the professional to a mission abroad is currently considered the main reason for the failure of the expatriate.

The literature points out several personal characteristics that need to be observed by the company, either when it internally chooses a professional to expatriate or it

S. Cristo-Andrade (✉)
PhD Student in Management at University of Beira Interior, Covilhã, Portugal
e-mail: silveli.andrade@ubi.pt

B. Felix · E. W. Mainardes
Departament of Management, FUCAPE Business School, Vitória/ES, Brazil
e-mail: emerson@fucape.br; bfelix@fucape.br

M. Peris-Ortiz et al. (eds.), *Knowledge, Innovation and Sustainable Development in Organizations*, Innovation, Technology, and Knowledge Management,
https://doi.org/10.1007/978-3-319-74881-8_9

opts to hire (Ones and Viswesvaran 1999; Remhof et al. 2014). Among the characteristics known as personality traits, or Big Five, the openness to change characteristic appears as an important factor of analysis because of its positive relation with other elements that are also considered essential to the success of an expatriate, such as autonomy, cross-cultural adaptation, creativity, and innovation (Caligiuri 2000; Judge and Zapata 2015).

Allied to the openness to change characteristic, another factor that also contributes to the success of the professional abroad, thus increasing the competitive advantage of an organization, is the so-called work engagement. This positive alignment between the professional and the organization, evidenced by Santos (2015), emphasizes that professionals with this feeling feel prepared and motivated for future tasks, which drive them to success.

In understanding the importance of this success of the expatriate to organizations and the great attention that the literature reserves for the subject (Araújo et al. 2012; Black et al. 1992; Chen and Lin 2014; Doherty & Freitas 2011; Engle et al. 2015; Kubo and Braga 2013; Santos 2015; Shaffer et al. 2012; Vance and Mcnulty 2014), we verify that little is known about the paradox surrounding this personality trait, openness to change, as well as the effects it can provide in social and organizational interactions.

Among the works that have analyzed this personality trait as an antecedent to the intention to be expatriated, we note that it has often been pointed out as desirable (Bobowik et al. 2011; Ones and Viswesvaran 1999; Remhof et al. 2014). However, some studies suggest a high risk of turnover of these same professionals who hold this promising characteristic (Cieciuch and Davidov 2012; Dickmann et al. 2008).

For this reason, we argue that the personality trait openness to change presents a paradoxical feature in expatriation candidates in an organization. We suggest that professionals with a high degree of openness to change tend to be simultaneously more likely to be expatriates as well as presenting a high propensity to leave the company. We also support that when the professional presents higher levels of work engagement, such a paradoxical relationship tends to attenuate.

The model that this study proposes is to identify the personality trait openness to change as an antecedent to both the intention to expatriate and the intention to leave the company, with work engagement having a moderating role in both relations.

9.2 The Personality Trait Openness to Change

It is no secret that the knowledge of human behavior has always been a challenge for psychology. One of the paths used to extend this knowledge focuses on testing the personality traits that each individual possesses (Judge and Zapata 2015). Personality traits distinguish how each individual behaves in a variety of situations, with two main points as principle: the stability of this trait in face of time and the influence it exerts on the behavior of individuals (Matthews et al. 2003).

In several previous studies (e.g., Araújo et al. 2012; Baruch and Altman 2002; Dickmann et al. 2008; Gallon et al. 2014; Loes 2015; Ones and Viswesvaran 1999; Remhof et al. 2014), personality traits were identified as important factors both in the intention of an individual working outside the country and in the performance that this expatriate employee presents in their international mission.

These personality traits selected by the literature, known as Big Five or five-factor model, identify the following personality characteristics of an individual: openness to change, neuroticism, extraversion, kindness, and conscientiousness.

Among these characteristics, this study focuses on the personality trait openness to change, which is because of its catalytic role in the intention of the professional who is applying for such a challenge, as well as the fact that this personal characteristic can act as a guide to the success or failure of this same professional (Gallon et al. 2014). We must recall that it is precisely this success or failure result of the professional that will directly impact a competitive advantage of the organization, especially in the dissemination of knowledge in international operations.

This personality trait openness to change can be defined as a personal characteristic of the individual in presenting skills such as creativity, innovation, high degree of independence, extraversion, and curiosity (Judge and Zapata 2015; Remhof et al. 2014).

In thinking of the international mission as a challenge, when a professional chooses to accept it, we need to emphasize that some discomfort, stress, or even frustration can occur. These symptoms are considered normal phenomena in scenarios with constant changes, i.e., in dynamic environments (Pereira et al. 2005).

This change of a professional from their domestic work base to an overseas work base generates tensions that tend to demand a greater ability of this professional to face the situation. They will need to deal not only with the new professional setting but with a range of adaptations, such as a new society, a new culture, a new routine, a new climate, a new language, and finally new norms either unknown or not experienced by them, which can lead to various situations of stress (Black et al. 1991; Bobowik et al. 2011; Caligiuri 2000).

It is at this point that Judge and Zapata (2015) inform that professionals who have this openness to change personality trait usually have a profile that is desired by organizations that want to conquer new markets by expatriating their professionals. This is because they have greater autonomy and they are creative and innovative, allowing greater independence to solve problems without the support of a direct team, as would happen in a domestic environment.

In another study, Caligiuri (2000) points out that this personality trait also provides the professional with a possibility of greater acceptance by the residents of the new host country, thus facilitating and motivating a rapid adjustment with the new environment.

Thus, when a professional is open to change, that is, open to new people, new cultures, and new environments, one tends to associate these new experiences positively, reducing the stress of this transition and facilitating their adjustment in the new scenario. We must emphasize that the cultural learning that openness to relationships generates only tends to contribute to the accomplishment of their new professional functions, reducing possible barriers that would be more difficult to transpose in the case of one isolating himself (Araújo et al. 2012; Caligiuri 2000; Thoresen et al. 2004).

Once the role that openness to change performs in an individual's social and professional skills is explained, we also need to understand what contains this relevant personal characteristic. Openness to change adds dimensions such as creativity, curiosity, and sophistication, dimensions that only tend to enhance the performance of a professional in dynamic scenarios (Judge and Zapata 2015; Remhof et al. 2014; Thoresen et al. 2004).

Thus, in short, people who have this personality trait tend to be imaginative and original, that is, creative, they do not stick to the traditional, and as it is popularly said, "they get out of the box"; by being curious they always seek learning, be it professional or intercultural, they explore and go beyond possible stipulated limits, and they act proactively and are tolerant; their sophistication always seeks the aesthetic, the beautiful, what inspires them, and what brings them pleasure, transforming themselves as well as the environment they are inserted in (Andrade 2008).

Even so, despite the importance that literature attaches to the personality trait, openness to change is essential for professionals who will work outside the country; Kulkarni et al. (2010) warn that for this action to succeed, then more is required in addition to this trait that it also presents other personal and professional skills to intensify this openness to change, such as technical competence, family bond, and good sociability (Baruch and Altman 2002; Loes 2015; Remhof et al. 2014).

9.3 Intention of the Professional to Be Expatriated by the Company

The intention of the professional to be expatriated by the company can be defined as a result of an acceptance of a proposal offered by employers for a work assignment outside the country or as a unilateral decision by the individual to migrate and work outside the country, even without any invitation from their employer, i.e., the individual plans his career where migration is one of the steps to be taken (Engle et al. 2015; Remhof et al. 2014).

Based on data from the United Nations, in its International Migrant Report 2016, migration around the world has only increased in recent decades. In 2005 there were 195 million emigrants scattered across the globe. There are now 244 million emigrants who are out of their countries for a variety of reasons, whether they are looking for a new life, refugees seeking safety, or professionals at work (United Nations 2016).

Regarding working professionals, the Breakthrough to the Future of Global Talent Mobility report recently published the result of a survey of 163 global organizations reporting that there are more than 11 million expatriate professionals,

54% of which are in the Americas; 40% in Europe, the Middle East, and Africa; and 6% in Asia (BGRS 2016). Organizations worldwide are placing their professionals to expand their business and disseminate knowledge.

Also referring to these working professionals, authors such as Shaffer et al. (2012) emphasize that although this growth remains in the current global scenario, there is still great difficulty in managing expatriate professionals, and companies have already considered alternatives to reduce costs in this sector. The inclusion of technology use such as online meetings or task monitoring, international missions within the space of a year, and short business trips is some of the possibilities where companies can afford a modest investment, compared to the investment of maintaining their employee for a long time outside the home country.

However, with this growing movement of professionals identified by both the UN and human resources companies, it is impossible not to acknowledge how important professional emigration to modern societies and economies has become. Therefore, by focusing on professional emigrants who are spread around the world, we question, what makes a professional want to be expatriated by an organization? What motivates them to emigrate to work in another country?

Researchers such as Remhof et al. (2014) warn in their study that care must be taken in handling the motivation that makes a professional choose to be expatriated. According to these authors, there is a difference between a professional having the willingness, a predisposition to be expatriated, and intending to be expatriated, i.e., will and intention need to be treated as different phenomena.

Willingness is considered as less cognitive, i.e., it is only a disposition that the professional shows if an opportunity appears. One does not pursue it but rather waits for it, being open to the possibility. Intention, in turn, one observes that it is a reasoned process, i.e., the individual organizes for an international mission, and it is a premeditated behavior thinking ahead of improvements, be it regarding their career or their abilities or the gain resulting from learning. He turns the mission into a goal, even without it being offered to them yet (Bassy 2002; Remhof et al. 2014).

In a similar vein, Santos (2015) comments that what leads a professional to want to be expatriated can also result from two sources: intrinsic motivation and extrinsic motivation. Intrinsic motivation is that which is part of the "self"; they are manifestations that have grown with the individual since childhood and they appear throughout life, i.e., it can be the curiosity for the world, the ease of accepting challenges, the quest for victory, or even the simple pleasure of performing some challenging activity. In the extrinsic, this motivation comes from outside the individual and can be a monetary reward, an external view of their work, an external view of the individual himself, or even an organizational pressure to accept the mission as the only option for progression in their career.

Another view on what makes a professional choose to be expatriated concludes that the main reasons can be first the personal challenge, followed by professional development, the career itself, the importance of the task, the opportunity for growth, the geographical location of the new work base, and the monetary reward (Spohr 2011; Shaffer et al. 2012).

In this case of motivation, we must mention that researchers such as Sagie et al. (1996) warn that motivation may be different for expatriates from different cultures. In dealing with uncertainties, risks, or difficulties, aiming at oneself's individual improvement may not be so attractive to professionals who experience cultures where the group prevails rather than the individual. In this case an extrinsic motivation, such as the monetary reward, may change position in the ranking of the main reasons for expatriation.

Although there are differences in how each professional behaves, one fact is certain: the motivation to be expatriated varies from an internal position of the individual or from an external stimulus to that individual. By associating then, the intention of being expatriated with the benefits that several authors (Andrade 2008; Baruch and Altman 2002; Caligiuri 2000; Gallon et al. 2014; Kulkarni et al. 2010; Thoresen et al. 2004) suggest that the openness to change personality trait propitiates to a professional, we constructed the first assumption for the proposed model, A1 – the personality trait openness to change is positively associated with the intention to be expatriated by a particular company.

9.4 Intention to Leave the Company

The reasons listed as defining a decision by the professional to leave their company after or during an international mission may be several. In the literature, there are studies that suggest that 10–25% of professionals returning from a mission abroad leave their jobs after a year of return (Black et al. 1992; Vidal et al. 2008). Other studies mention that this loss may exceed 30% of the total number of professionals expatriated by the organization (Baruch and Altman 2002; González and Oliveira 2011; Loes 2015; Spohr 2011).

Regardless of the percentage that it reaches, the loss of this professional is pointed out countless times as a critical point of the organization, being necessary to note that in this fact lies the "Achilles heel" of multinational organizations, as observed by Baruch and Altman (2002). Overcoming this difficulty may be paramount for the company to achieve a competitive advantage in the global marketplace.

Complementing the importance of this organizational concern, it is fact that it is not only the financial investment that is lost with the departure of the company professional. When an international professional leaves the organization, he/she also takes with him/her the knowledge and skills acquired, his/her leadership, as well as the professional and social network that he/she created in this period working for the organization (Bolino 2007).

Starting from the positioning of a professional who accepts an international mission, what makes this same professional at some point in that experience choose to leave the company?

In this case, there may be several factors that guide the professional that even by accepting such a mission, with the course of time, chooses to leave the company.

When the professional requests resignation during their international mission, the factors can be problems of adaptation to the new country, difficulties in the new position, or even family matters that result in them wanting to return to their home country (Loes 2015; Pereira et al. 2005).

However, when this request for dismissal happens back in the home country, it may be due to a poor repatriation by the organization, problems in career progression, a feeling of isolation, loss of autonomy, lack of valorization, or struggles in financial matters. All these factors can cause professional dissatisfaction, leading one to choose to leave the company even having succeeded in their international mission (Bolino 2007; Herman and Tetrick 2009; Stahl et al. 2009; Vidal et al. 2008).

Although "dissatisfaction" is the first thought when a professional chooses to leave an organization after returning to the country of origin, this aspect raises another aspect for such a decision, the personality trait "openness to change."

This personality trait is classified as a gift in positions with international missions; nonetheless, people who have it are also considered to be people with a high degree of autonomy, prone to a continuous search for overcoming, hungry for the new, and consequently search for new challenges when old ones have already been overcome (Judge and Zapata 2015).

Another way to sharpen the intention to leave the company of the professional is the invitations they receive from other organizations. Professionals who have these characteristics and personal skills already cited, added to the international learning they now incorporate, become highly attractive to headhunters. These organizational agents often seek this internationally prepared professional to relocate them to new organizations (Shen and Hall 2009). This possible migration by the professional from an organization to another can result in a loss of investment and knowledge in this professional.

The intention to leave the company can also, in addition to all these possibilities already mentioned, be an internal feeling of the professional, i.e., due to their positioning being directed to the international learning and not to the demand of the organization (Shaffer et al. 2012). With his desire being to maintain their growth in the global career, the professional demonstrates that "his career planning" is above the planning that the organization intends for them. In these cases the professional makes the decision to leave the company even before the return, already looking for or accepting offers from headhunters or other organizations with which they already have interaction (Stahl et al. 2009).

Based on this principle, which professionals with this personality trait openness to change are challenging, confident, decisive, and autonomous people regarding their career and their life planning (Caligiuri 2000; Remhof et al. 2014; Santos 2015), we constructed the second assumption for the proposed model. It presents precisely this decisive and autonomous character of an individual in face of their career, facing the choice between staying in a company and seeking a new placement in another company in which they are interested, A2 – the personality trait openness to change is positively associated with the intention to leave a particular company.

9.5 The Role of Work Engagement

Santos (2015) defines that work engagement is an alignment between the professional and the organization, but this alignment is not just an informal contract between employee and employer; it is more than that; it is a positive feeling of commitment and motivation that the individual demonstrates toward the organization. The professional takes for himself the values that the organization cherishes, thus putting all their energy, so that together with the organization they reach the established goals.

Several authors (Bakker and Bal 2010; Bakker and Demerouti 2008; Bhatnagar 2012; Schaufeli and Bakker 2004; Schaufeli et al. 2006) assert that the engagement that the professionals demonstrate by the organization is presented in three dimensions: vigor, dedication, and absorption. These dimensions together reflect on the behavior and commitment that the professional demonstrates toward the organization.

On these dimensions, it can be said that vigor is a great positive energy that the professional imputes in his work and does so without great physical or psychological wearing out; the dedication demonstrates how much this professional is involved with the organization and how proud and enthusiastic they are due to work; and finally, absorption is the feeling of how "immersed" this individual is in his activities, i.e., the professional committing himself completely for a mission in which they believe and feel good in performing (Medeiros et al. 2005; Schaufeli et al. 2002; Schaufeli and Bakker 2003).

Besides the effect of these dimensions on the individual, there are some resources that can lead the professional to this engagement. In the literature, the so-called work resources are considered factors that can motivate professional engagement but also soften possible negative relations that arise. Some of these resources may be the autonomy that the professional has at work, the feedback of the performance that the company provides, the social support of the team and the family, or even the opportunities offered by the company to learn and develop their skills (Schaufeli and Bakker 2004).

Personal resources are extracted from the behavior of the individual. We consider self-evaluation, self-esteem, positivity, and resilience, in short, characteristics that make the professional shape the environment in which they are when necessary, creating balance to themselves and to those who surround them (Bakker and Demerouti 2008). These personal resources of the individual arise from the passion for what one does, from their dedication to what they believe, and of the efficiency and effectiveness in transforming disturbed behavior and environments into positive attitudes and locations (Bakker and Bal 2010; Qadeer et al. 2016; Santos 2015).

With these work and personal resources at hand, engaged professionals are those who believe in themselves and are active, open, inclusive, and optimistic, energizing everything around them, generating their own positive feedback, committing themselves to the organization, and assuming the values for themselves (Schaufeli et al. 2002).

Loaded with this positive attitude, professionals open to change do not differ much from this behavior. They are also filled with this energy and characteristics that strengthen them in new environments. They do not allow themselves to be overwhelmed by unexpected events and always try to modify their own behavior to ease any difficult situation. Their ability to deal with the new open doors widens their network (Caligiuri 2000).

Thus, professionals who feel engaged at work and are open to change seek their improvement; they seek challenges (Remhof et al. 2014). Due to this predisposition presented by professionals engaged in the work and imbued with this openness to change personality trait, we constructed the third assumption for the proposed model: A3a – Work engagement moderates the relationship between the personality trait openness to change and the intention to be expatriated by a particular company, so that this relationship becomes stronger when work engagement is higher.

After the presentation of the third assumption, which focused on the moderating role of the work engagement construct, we need to address the relationship that this construct presents when related to the intention to leave the company. Previous studies provide evidence that professionals who feel engaged at work tend to show a low predisposition to leave the company (Bakker and Bal 2010; Bhatnagar 2012; Bolino 2007; Stahl et al. 2009). In this view, we assume that work engagement moderates the intention to leave the company of an individual openness to change, and we thus constructed the last assumption for the proposed model: A3b – Work engagement moderates the relationship between the personality trait openness to change and the intention to leave the company, so that this relationship becomes weaker when work engagement is higher.

9.6 The Proposed Model

To test the relationship between openness to change, intention to be expatriated, intention to leave the company, as well as the moderating role of the work engagement construct conducted in the two relations that the personality trait openness to the change can act, we constructed the model according to Fig. 9.1.

What we propose in this model is to test whether work engagement acts when it focuses on relationships where individuals are open to change. In the literature, professionals with this personality trait are considered prone to accept an expatriation by their companies (Baruch and Altman 2002; Dickmann et al. 2008; Gallon et al. 2014; Loes 2015), as well as being considered individuals with a greater degree of autonomy regarding their careers (Shaffer et al. 2012; Stahl et al. 2009). However, the interest of the model goes beyond testing these assumptions; we can include in this equation the possibility of moderation that the work engagement construct can provide in these relations.

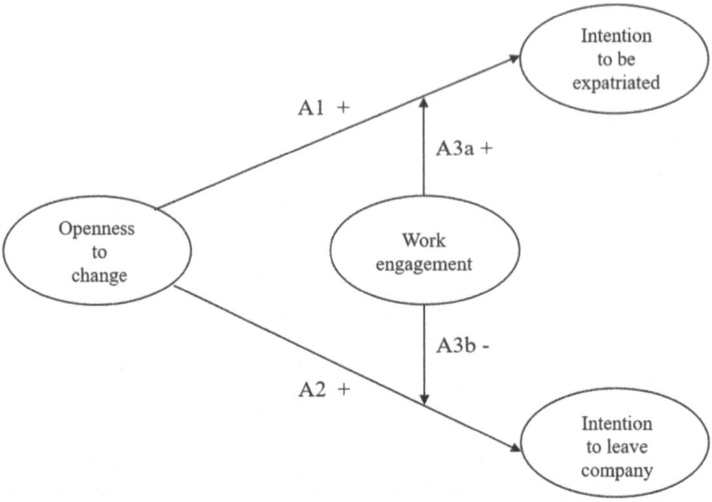

Fig. 9.1 Proposed model. A1 – The personality trait openness to change is positively associated with the intention to be expatriated by a particular company. A2 – The personality trait openness to change is positively associated with the intention to leave a particular company. A3a – Work engagement at moderates the relationship between the personality trait openness to change and the intention to be expatriated by a particular company, so that this relationship becomes stronger when work engagement is higher. A3b – Work engagement at moderates the relationship between personality trait openness to change and intention to leave a particular company, so that this relationship becomes weaker when work engagement is higher

9.7 Final Considerations

Nowadays it is relevant for organizations to know a little more about this paradox that the personality trait openness to change can create, when a professional with this characteristic so appreciated by organizations is placed in a dynamic and challenging scenario.

We expect that this personality trait openness to change will act positively in candidates for expatriation, as well as that this same trait may be present in professionals who do not feel trapped in their current positions and have greater autonomy to change companies. It is precisely this paradox that becomes important to understand whether it is desirable or not for organizations seeking a competitive advantage in the global marketplace to have, in its framework of international actors, professionals who hold the personality trait openness to change.

It is well known that more than a quarter of professionals working on foreign missions face difficulties in the culture of the home country, due to lack of skills or characteristics necessary to face the new environment. For this reason, hiring, retaining, and motivating professionals for expatriation have been pointed out as an important point in global mobility management within organizations that seek success in their foreign missions.

With the possibility of measuring a context with the proposed model, one can achieve results that increase the knowledge about how these professionals that act internationally behave, thus directing hiring in order to achieve the expected success of these professionals.

The selection of expatriate candidates, where we evaluate not only their personality traits but also their engagement at work, can offer greater assertiveness, helping companies to reduce risks that may arise, such as failure in the foreign mission and loss of knowledge acquired of capital and human investment during the international mission.

Expanding knowledge on the moderating role that work engagement can provide in the relationship of the openness to change trait with the intentions of expatriating or leaving the company is an important point of this study. And in addition to this contribution, we present the possibility of a greater understanding of the paradox that this personality trait, openness to change, can present in the proposed scenario.

We also suggest some new lines of research, for example, a comparison of the behavior regarding the intention to expatriate between the genders, since a disparity of applicants for expatriation in some cultures has already been observed.

Another suggestion for future research would be a comparison between professionals who master foreign languages and those who do not master it, since the consideration of this control data could provide new results to the study. Finally, the degree of internationalization of respondents' companies could also be added to the control data, improving the visualization of the real scenario that the professional is in.

9.8 Acknowledgments

This research was supported by Brazilian National Council for Scientific and Technological Development (CNPq/Brazil – project 303669/2015–2), FAPES (Fundação de Amparo à Pesquisa e Inovação do Espirito Santo/Brazil), and Portuguese Science Foundation (FCT) through NECE – Núcleo de Investigação em Ciências Empresariais (Programa de Financiamento Plurianual das Unidades de I&D da FCT – Fundação para a Ciência e Tecnologia, Ministério da Ciência, Tecnologia e Ensino Superior/Portugal – UID/GES/04630/2013).

References

Andrade, J. M. (2008). *Evidências de validade do inventário dos cinco grandes fatores de personalidade para o Brasil. Tese de doutorado, Psicologia Social, do Trabalho e das Organizações*. Brasil: Universidade de Brasília, Brasília.

Araújo, B. F. V. B., Bilsky, W., & Moreira, L. M. C. O. (2012). Valores pessoais como antecedentes da adaptação transcultural de expatriados. *RAM: Revista de Administração Mackenzie, 13*(3), 69–95.

Araújo, B. F. V. B., Texeira, M. L. M., Cruz, P. B., & Malini, E. (2012). Adaptação de expatriados organizacionais e voluntários: similaridades e diferenças no contexto brasileiro. *Revista de Administração da Universidade de São Paulo, 47*(4), 555–570.

Bakker, A. B., & Bal, M. P. (2010). Weekly work engagement and performance: A study among starting teachers. *Journal of Occupational and Organizational Psychology, 83*(1), 189–206.

Bakker, A. B., & Demerouti, E. (2008). Towards a model of work engagement. *Career development international, 13*(3), 209–223.

Baruch, Y., & Altman, Y. (2002). Expatriation and repatriation in MNCs: A taxonomy. *Human Resource Management, 41*(2), 239–259.

Bassy, M. (2002). *Motivation and work-Investigation and analysis of motivation factors at work.* PhD Thesis, Department of Management and Economics, Linköping University, Swedish. Retrieved Nov 12, 2016, from http://www.divaportal.org/smash/get/diva2:17451/ fulltext01

BGRS - Brookfield Global Relocation Services. (2016). *Global mobility trends. USA: Brookfield global relocation services*. Retrieved Nov 10, 2016, from http://globalmobilitytrends.brookfieldgrs.com/

Bhatnagar, J. (2012). Management of innovation: Role of psychological empowerment, work engagement and turnover intention in the Indian context. *The International Journal of Human Resource Management, 23*(5), 928–951.

Black, J. S., Gregersen, H. B., & Mendenhall, M. E. (1992). Toward a theoretical framework of repatriation adjustment. *Journal of International Business Studies, 23*(4), 737–760.

Black, J. S., Mendenhall, M., & Oddou, G. (1991). Toward a comprehensive model of international adjustment: An integration of multiple theoretical perspectives. *Academy of Management Review, 16*(2), 291–317.

Bobowik, M., Basabe, N., Páez, D., Jiménez, A., & Bilbao, M. A. (2011). Personal values and Well-being among Europeans, Spanish natives and immigrants to Spain: Does the culture matter? *Journal of Happiness Studies, 12*(3), 401–419.

Bolino, M. C. (2007). Expatriate assignments and intra-organizational career success: Implications for individuals and organizations. *Journal of International Business Studies, 38*(5), 819–835.

Caligiuri, P. M. (2000). Selecting expatriates for personality characteristics: A moderating effect of personality on the relationship between host national contact and cross-cultural adjustment. *MIR: Management International Review, 40*(1), 61–80.

Caligiuri, P. M. & Lazarova, M. (2001). Strategic repatriation policies to enhance global leadership development. Developing global business leaders: Policies, processes, and innovations (pp. 243–256). Retrieved Oct 23, 2016, from https://www.researchgate.net/profile/Paula_Caligiuri/publication/234021410_Strategic_Repatriation_Policies_to_Enhance_Global_Leadership_Development/links/56321fc908ae3de9381ef89b.pdf

Chen, J.-H., & Lin, J.-Z. (2014). Prediction rules of the employees. *Journal of Marine Science and Technology, 22*(2), 116–124.

Cieciuch, J., & Davidov, E. (2012). A comparison of the invariance properties of the PVQ-40 and the PVQ-21 to measure human values across German and Polish Samples. *Survey Research Methods, 6*(1), 37–48.

Dickmann, M., Doherty, N., Mills, T., & Brewster, C. (2008). Why do they go? Individual and corporate perspectives on the factors influencing the decision to accept an international assignment. *The International Journal of Human Resource Management, 19*(4), 731–751.

Engle, R. L., Schlägel, C., Dimitriadi, N., Tatoglu, E., & Ljubica, J. (2015). The intention to become an expatriate: A multinational application of the theory of planned behaviour. *European Journal of International Management, 9*(1), 108–137.

Freitas, M. E. (2011). Expatriação Profissional: o desafio interdependente para empresas e indivíduos. *Gestão e Sociedade, 4*(9), 689–708.

Gallon, S., Scheffer, A. B. B., Bitencourt, B. M. & Gallon, I. (2014). *Processo de expatriação: além da experiência internacional. Gestão Contemporânea, 16(july/dec)*. Retrieved Oct 20, 2016, from http://seer4.fapa.com.br/index.php/arquivo/article/viewFile/355 /161

González, J. M. R., & Oliveira, J. A. (2011). Os efeitos da expatriação sobre a identidade: estudo de caso. *Cadernos EBAPE. BR, 9*(4), 1122–1135.

Herman, J. L., & Tetrick, L. E. (2009). Problem-focused versus emotion-focused coping strategies and repatriation adjustment. *Human Resource Management, 48*(1), 69–88.

Judge, T. A., & Zapata, C. P. (2015). The person–situation debate revisited: Effect of situation strength and trait activation on the validity of the Big Five personality traits in predicting job performance. *Academy of Management Journal, 58*(4), 1149–1179.

Kubo, E. K. M., & Braga, B. M. (2013). Ajustamento intercultural de executivos japoneses expatriados no Brasil: um estudo empírico. *RAE-Revista de Administração de Empresas, 53*(3), 243–255.

Kulkarni, M., Lengnick-Hall, M. L., & Valk, R. (2010). Employee perceptions of repatriation in an emerging economy: The Indian experience. *Human Resource Management, 49*(3), 531–548.

Loes, C. (2015). Improving processes for expatriate selection. *Business Studies Journal, 7*(2), 99–112.

Matthews, G., Deary, I. J., & Whiteman, M. C. (2003). *Personality traits* (2nd ed.). Cambridge: Cambridge University Press. Retrieved Feb 16, 2017, from http://content.schweitzeronline.de/static/catalog_manager/live/media_files/representation/zd_std_orig__zd_schw_orig/002/194/905/9780521831079_content_pdf_1.pdf

Medeiros, C. A. F., Albuquerque, L. G., Marques, G. M., & Siqueira, M. (2005). Um estudo exploratório dos múltiplos componentes do comprometimento organizacional (43a ed.). *REAd: Revista Eletrônica de Administração, 11*(1), 1–22.

Ones, D. S., & Viswesvaran, C. (1999). Relative importance of personality dimensions for expatriate selection: A policy capturing study. *Human Performance, 12*(3–4), 275–294.

Pereira, N. A. F., Pimentel, R., & Kato, H. T. (2005). Expatriação e estratégia internacional: o papel da família como fator de equilíbrio na adaptação do expatriado. *RAC-Revista de Administração Contemporânea, 9*(4), 53–71.

Qadeer, F., Ahmad, A., Hameed, I., & Mahmood, S. (2016). Linking passion to organizational citizenship behavior and employee performance: The mediating role of work engagement. *Pakistan Journal of Commerce and Social Sciences, 10*(2), 316–334.

Remhof, S., Gunkel, M., & Schlägel, C. (2014). Goodbye Germany! The influence of personality and cognitive factors on the intention to work abroad. *The International Journal of Human Resource Management, 25*(16), 2319–2343.

Sagie, A., Elizur, D., & Yamauchi, H. (1996). The structure and strength of achievement motivation: A cross-cultural comparison. *Journal of Organizational Behavior, 17*(5), 431–444.

Santos, G. A. (2015). *A motivação dos expatriados e o seu bem-estar no trabalho: estudo exploratório numa empresa de construção em África. Dissertação de mestrado Integrado em Psicologia*. Lisboa: Universidade de Lisboa.

Schaufeli, W. B. & Bakker, A. B. (2003). *UWES – Utrecht Work Engagement Scale. Preliminary manual. Occupational health psychology unit*. Utrecht University. Retrieved Nov 12, 2016, from http://www.wilmarschaufeli.nl/publications/Schaufeli/Test%20Manuals/Test_manual_UWES_Brazil.pdf

Schaufeli, W. B., & Bakker, A. B. (2004). Job demands, job resources, and their relationship with burnout and engagement: A multi-sample study. *Journal of Organizational Behavior, 25*(3), 293–315.

Schaufeli, W. B., Bakker, A. B., & Salanova, M. (2006). The measurement of work engagement with a short questionnaire a cross-national study. *Educational and Psychological Measurement, 66*(4), 701–716.

Schaufeli, W. B., Salanova, M., González-Romá, V., & Bakker, A. B. (2002). The measurement of engagement and burnout: A two sample confirmatory factor analytic approach. *Journal of Happiness Studies, 3*(1), 71–92.

Shaffer, M. A., Kraimer, M. L., Chen, Y. P., & Bolino, M. C. (2012). Choices, challenges, and career consequences of global work experiences: A review and future agenda. *Journal of Management, 38*(4), 1282–1327.

Shen, Y., & Hall, D. T. T. (2009). When expatriates explore other options: Retaining talent through greater job embeddedness and repatriation adjustment. *Human Resource Management, 48*(5), 793–816.

Spohr, N. (2011). *O processo repatriação na visão de profissionais repatriados brasileiros. Dissertação de mestrado em Administração de Empresas*. São Paulo: Fundação Getúlio Vargas.

Stahl, G. K., Chua, C. H., Caligiuri, P., Cerdin, J. L., & Taniguchi, M. (2009). Predictors of turnover intentions in learning-driven and demand-driven international assignments: The role of repatriation concerns, satisfaction with company support, and perceived career advancement opportunities. *Human Resource Management, 48*(1), 89–109.

Tharenou, P., & Caulfield, N. (2010). Will I stay or will I go? Explaining repatriation by self-initiated expatriates. *Academy of Management Journal, 53*(5), 1009–1028.

Thoresen, C., Bradley, J. C., Bliese, P. D., & Thoresen, J. D. (2004). The big five personality traits and individual job performance growth trajectories in maintenance and transitional job stages. *Journal of Applied Psychology, 89*(5), 835–853.

United Nations. (2016). *International migration report 2015*. Department of Economic and Social Affairs, Population Division. Retrieved Nov 12, 2016, from http://www.un.org/en/development/desa/population/migration/publications/migrationreport/docs/MigrationReport2015.pdf

Vance, C. M., & Mcnulty, Y. (2014). Why and how women and men acquire global career experience: A study of American expatriates in Europe. *International Studies of Management & Organization, 44*(2), 34–54.

Vianna, N. P., & Souza, Y. S. (2009). Uma análise sobre os processos de expatriação e repatriação em organizações brasileiras. *Revista Base Administração e Contabilidade da UNISINOS, 6*(4), 340–353.

Vidal, M. E. S., Valle, R. S., & Aragón, M. I. B. (2008). International workers' satisfaction with the repatriation process. *The International Journal of Human Resource Management, 19*(9), 1683–1702.

Chapter 10
The Entrepreneurial University Stimulating Innovation Through Campus Development: The MIT Case

Flavia T. J. Curvelo Magdaniel

10.1 Introduction

Universities are considered organizations operating in environments of rapid change. Over the last few decades, a common trend from traditional missions of teaching and research to the third mission of economic development has been observed in universities Europe and America (Sam and Van Der Sijde 2014). The concept of entrepreneurial university positions them as new economic actors emphasizing their collaboration with external stakeholders for socioeconomic development (Etzkowitz 2004). Entrepreneurial universities are leading in education, advancing research, controlling their resources, organizing their own capacity to transfer technologies, and fostering entrepreneurship as culture among their faculty and students (Drucker and Goldstein 2007; Etzkowitz 2008; Vorley and Nelles 2008). Similarly, there are raising concerns about the entrepreneurial roles of universities in the so-called academic capitalism. This phenomenon is defined as a wide variety of market (and market related) activities used by faculty and institutions to secure external funding due to reduced public funding such as patenting, spin-off companies, grants, university-industry partnerships, and tuition fees (Kauppinen 2012). According to Jessop (2017), entrepreneurialism in universities has a longer history than academic capitalism as suggested by Schumpeter's ideas on innovation and discussed in earlier research (Deem 2001; Marginson and Considine 2000). Undoubtedly, the evolutionary role of universities from traditional to entrepreneurial is linked to transformations in the dynamic context in which they operate.

F. T. J. Curvelo Magdaniel (✉)
Department of Management in the Built Environment, Faculty of Architecture, The Built Environment, Delft, The Netherlands

The Built Environment, Delft University of Technology, Delft, The Netherlands
e-mail: f.t.j.curvelomagdaniel@tudelft.nl

M. Peris-Ortiz et al. (eds.), *Knowledge, Innovation and Sustainable Development in Organizations*, Innovation, Technology, and Knowledge Management,
https://doi.org/10.1007/978-3-319-74881-8_10

The knowledge-based economy (KBE) has influenced the evolutionary path of universities worldwide. According to (Cooke and Leydesdorff 2006), the KBE is regarded as a system used by governments to frame their perspectives for developing science, technology, and innovation policies. Herein, the basic idea of knowledge as an economic factor attributed to Schumpeter (1934) has developed by changing the composition of the labor force as well as technological and institutional trajectories. In the KBE, the range of technology-based research activities has increased, both in number and related processes, with the advancements of technologies during the ICT industrial revolution and the digital and information revolution (Headrick 2009). Many companies in developed economies invest more on R&D and are increasingly engaged in these activities with universities. Correspondingly, the number of people employed in research is growing steady in many countries (OECD 2013). With globalization and the changing dynamics of mobility patterns, most of the competitive advantage of countries and organizations relies on their ability of attracting and retaining talented people. Universities do compete on this basis.

There are different ways in which universities must sustain their competitive advantage in the KBE. In education and research, competition is understood and measured in nonmonetary terms such as prestige, recognition, or distinction (Schulze-Cleven et al. 2017). Academic reputation becomes a major factor influencing students' university choice (Briggs 2006). Financial competitiveness in universities is shaped by marketization and results from the increased numbers of students and faculty worldwide and the shifting role of academic capitalism where public and private expenditures are intertwined (Schulze-Cleven and Olson 2017). Herein, the cuts in public funding are pressing universities to diversify their external funding sources and raise revenues through tuition fees, donations, business-university cooperation, and returns on their endowments. Conversely, the cost of maintaining research universities is growing since "world-class research" demands adequate facilities and access to appropriate libraries, laboratories, offices, internet, and other resources (Altbach 2004). In a context where obtaining and sustaining financial resources is increasingly competitive, universities must use them efficiently also to embrace their evolutionary roles.

The campus can be seen as one of these strategic resources. Investing in physical infrastructure has already been addressed as a way to strengthen the relationships between universities, industry, and governments (Van Winden 2008). This infrastructure is regarded in early global policies as part of national science systems (OECD 1996). Facilities, transportation networks, and telecommunication systems have been outlined as enablers of economic activities at national level (Porter 1990). In organizations, Joroff (1993) emphasized the view of managing real estate as the fifth resource besides human resources, technology, capital, and information technologies. Existing studies in university campus management outline the enabling function of real estate in attaining competitive advantage in the KBE (Curvelo Magdaniel 2016; Den Heijer 2011). This chapter takes forward this perspective by illustrating how a leading entrepreneurial university has used its campus as strategic resource to react timely and adequately to the dynamic context of the KBE. The first section links the concepts of dynamic capabilities and real estate strategic management to provide a conceptual framework. The second section describes the methodology used in this chapter. The third and the fourth sections describe and discuss the results, respectively. And the fifth section draws this chapter's main conclusions.

10.2 Dynamic Capabilities and Real Estate Strategic Management

The concepts of "dynamic capability" and "real estate strategic management" are linked through the fundamental question in the field of strategic management, which according to Rumelt et al. (1994) lies on how organizations achieve sustainable competitive advantage. Dynamic capabilities are defined as a company's strategic ability to combine inside and outside competences to address volatile environments and periods of rapid change (Teece et al. 1997). These capabilities involve combinations of organizational, functional, and technological skills used to create, extend, and modify resources. From a resource-based perspective, firm performance depends on the firm's control of unique and difficult-to-imitate resources (Rumelt 1984; Teece 1984). These can be regarded as tangible assets (e.g., plant, equipment, raw materials, etc.) and intangible assets (e.g., tacit knowledge and relationships among actors). Srivastava et al. (1998) argue that the latter can give organizations a more sustainable form of competitive advantage since they are more difficult to imitate. Overall, the emphasis on management capabilities integrates and draws attention upon research in other areas such as management of R&D, product and process development, technology transfer, human resources, organizational learning, etc. (Teece et al. 1997). Similarly, corporate real estate management (CREM) fits in the dynamic capability approach to understand multiple and new sources of competitive advantage.

CREM is understood as "the management of a corporation's real estate portfolio by aligning the portfolio and services to the needs of the core business (processes), in order to obtain maximum added value for the business and to contribute optimally to the overall performance of the corporation" (Dewulf et al. 2000). In this field, corporate real estate (CRE) is regarded as "the fifth resource" (Joroff 1993). Herein, real estate is outlined as a facilitator of the primary processes of a corporation next to capital, human resources, information, and technology. This approach outlined the changing role of CRE in evolutionary stages from a "technical" to a "strategic" focus. Accordingly, the "alignment" between corporate and real estate strategies is central as well as the dynamic environment in which organizations operate. Too et al. (2010) argued that it is reasonable to see CRE as a source of capability to give companies its competitive advantage in a hypercompetitive climate. Although CRE can be considered a tangible asset, its strategic management is rather seen as an intangible asset. Therefore, CRE practices are harder for competitors to observe and imitate (Too et al. 2010). Managing CRE can be linked to the concept of orchestration as one of the means of gaining competitive advantage from the dynamic capabilities approach. Asset orchestration refers to the "managerial search, selection and configuration of resources and capabilities" (Helfat et al. 2009). In explicating dynamic capabilities, Teece (2007) disaggregates this concept into "the capacity (1) to sense and shape opportunities, (2) to seize opportunities, and (3) to maintain competitiveness trough enhancing, combining, protecting and when necessary, reconfiguring the business enterprise's intangible and tangible assets." From a CREM perspective, these capacities are expected to be deployed in the CRE practice.

The "articulation" between real estate strategy and corporate business strategy is a precondition to make effective real estate decisions favoring an enterprise's business (Nourse and Roulac 1993). This work pointed out that in obtaining such results, CRE managers must explicitly address how real estate strategies support corporate strategies. Furthermore, this study outlines that the driving force(s) of a company (in terms of products/markets, capabilities, and results) determines the business direction of a company, which changes over time with changes in specific environments. Hence, the dynamic environment in which organizations operate is an influential context for alignment between corporate and real estate strategy. This context and the particular culture and value of the organizations determine the appropriate real estate strategies that effectively support the broad business objectives of such organizations.

Today, the concept of dynamic capabilities has achieved a new relevance since the emerging technological landscapes enable firms to put dynamic capabilities into practice easier than even before (Shuen and Sieber 2009). Herein, they discussed the potentials of dynamic capabilities for web-enabled businesses affected by rapid technological change. These companies and other tech-driven organizations like universities have a similar driving force in attaining competitive advantage, i.e., the creation of new knowledge and its application to develop new technologies (Curvelo Magdaniel 2016). However, they have different values and culture influenced by their own distinct profiles. For instance, firms and universities differ because the former advance technologies mainly yield profit while the latter do so to advance science. Simultaneously, traditional and entrepreneurial universities differ in culture since the traditional mission of universities is limited to "educate people and advance research for society," while entrepreneurial universities add to this mission "to advance economic development." This difference in culture is an example of how some universities have changed their operations and their developing resources, including real estate as a reaction to the changing context of the KBE. The transition between traditional and entrepreneurial universities and the use of their real estate as a resource to attain sustainable competitive advantage are the focus of this chapter (Fig. 10.1).

Both dynamic capabilities and CREM emphasize the key role of strategic management in adapting, integrating, and reconfiguring particular competences to match the requirements of the changing environments. Identifying difficult-to-imitate competences is challenging for firms, and choosing domains of competences is sometimes influenced by past choices (Teece et al. 1997). This firms' continuation along a given trajectory is known as path dependency (Simmie 2005).

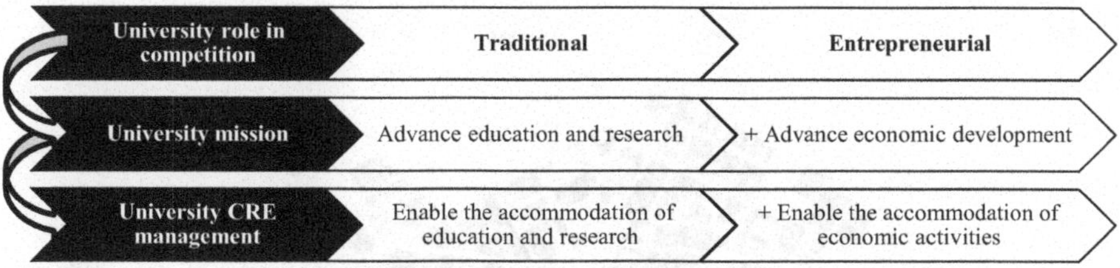

Fig. 10.1 Conceptual framework positioning CRE management as a strategic resource supporting universities' missions according to their changing roles

The companies' long-term commitments to certain domains of competence are quasi-irreversible and can lead to either growth or decline. Thus, dynamic capabilities and CREM are also shaped by the firms' evolutionary paths. This notion recognizes that history matters influencing the current position of the firm and their paths ahead. In this constrained context, the role of the managers in recognizing the available opportunities for their companies and orchestrating their resources to act upon them is crucial. Largely, balancing between short- and long-term decisions becomes critical to attain sustainable competitive advantage. This is especially important for CRE managers since they deal with tangible and static assets, which have long life cycles.

10.3 Methods

This chapter uses single case study as the main strategy (Flyvbjerg 2006) to illustrate how a leading entrepreneurial university has used its campus as strategic resource to react timely and adequately to the dynamic context of the KBE. It will illustrate, with rich and anecdotal descriptions, the CRE practice of an organization in a particular changing context. This chapter refers to some notions of dynamic capabilities addressed in the previous sections to provide an analytical perspective of this university's capacity to sense and seize its opportunities in crucial times while using the campus as strategic resource to maintain competitive advantage. This chapter uses data and information of a doctoral research (Curvelo Magdaniel 2016) carried out at Delft University of Technology.

10.3.1 Case Selection

The CRE practice studied in this chapter is the campus development of the Massachusetts Institute of Technology (MIT), which has become a role model of an entrepreneurial research university and has forged educational and research collaborations with universities, governments, and companies all over the world (Curvelo Magdaniel 2016; Etzkowitz 2008; Simha 2005). The MIT is a private nonprofit

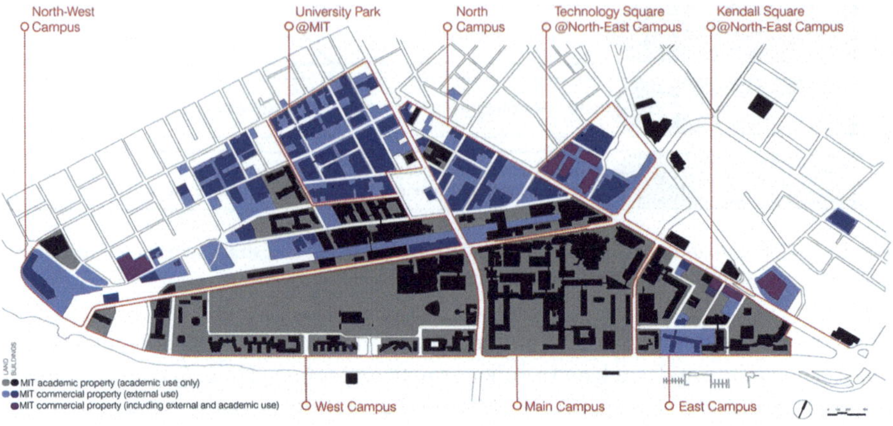

Fig. 10.2 MIT Campus in two types of properties and seven main development or planning zones (Curvelo Magdaniel 2016)

institution founded in Boston in 1861 and relocated to Cambridge in 1916. Since its foundation, the MIT emphasizes the "learning-by-doing" model, which was inspired in the typical education of the polytechnic universities that emerged in Europe at the end of the eighteenth century.

The MIT campus comprises about 104 hectares of land owned and leased by the MIT in Cambridge, which distinguishes two types of properties: the academic plant and the MIT's commercial real estate property. The first type accommodates only academic-related activities and therefore is a tax-exempt property. The second type is considered as a group of assets owned by the institute to generate income, adding to its financial resources. These properties are integrated in the urban fabric of the city and accommodate different activities including academic-related activities, R&D, housing, retail, and business (see Fig. 10.2).

10.3.2 Data Collection

This chapter used a variety of data sources for triangulation with the aim to document the CRE practice of campus development as a long-term process. The data was collected during the period from September 2014 to December 2015.

Open and semi-structured interviews with experts and key informants provided insight which lead to facts and relevant readings on the case and its context. Campus development experts possess an in-depth knowledge on particular domains of the CRE practice since they have been involved in this process over long periods (e.g., designers, planners, and managers with over 10 years of experience in the case). Three experts were contacted via e-mail and two of them responded (i.e., the campus planner during the period 1960–2000 and the campus real estate manager since 2000). These interviews include about 10–15 open questions to gain knowledge on the campus development process from each expert's experience. The interviews focused on the university's goals that have influenced the CRE practice, the implementation of CRE decisions, and to which extent CRE helped attaining such goals. Other experts had particular knowledge about contextual dynamics in the region. Two experts were contacted and responded via e-mail (i.e., two senior researchers on technology-driven real estate). These interviews include about 10 open questions aimed to gain insight into particular the contexts influencing each campus development. The interviews focused on the perceived relationship between innovation and CRE and the external developments influencing the campus development.

Key informants played a role leading to facts on campus development and extra insights on context-related information. They were contacted incrementally as suggested by experts or as indicated in reports while documenting the case. Fourteen key informants were contacted via e-mail, and 10 of them responded. This group included professionals in urban planning, real estate management, facility management, innovation policy, and entrepreneurship. These interviews were tailored inquiries on particular campus development phases, strategies, or decisions. They focused on campus development history in general.

Documentation allowed collecting the exact information containing references, names, and details of campus development covering a long time. This included maps and photos from archives, official briefings, administrative reports, existing empirical research on the cases, and articles in the media. The work of Simha (2001) significantly contributes to document of the CRE practice at MIT.

Site observations allowed insights into cultural features of the concepts and opinions and lead to facts. These were possible during site visits including field trips, seminar attendances, guided walks, and informal meetings with campus users.

10.3.3 Data Analysis

For analytical purposes, the notions in the conceptual framework served as instrument to generate insights. First, it was used to provide understanding of the MIT's path to establish its current position as a role model of entrepreneurial university. Second, it served to identify the CRE practices that supported this university's evolutionary roles from traditional to entrepreneurial. An iterative analytical procedure is used to sharpen the constructs by displaying enough evidence with examples, anecdotal reports, and descriptions. Mapping using open access applications (e.g., Google Earth, Esri Maps, Google Maps) allowed corroborating exact and particular information containing physical details on campus development over time. Overall, tying the emergent insights to the concepts from the literature also enhanced the processing and synthesis of the descriptive information.

10.4 Results

10.4.1 The MIT's Path to an Entrepreneurial University

Massachusetts has been a center of industrial and technological innovation for a long time. In the last 70 years, it has successfully adapted technological changes from electronics to software and to biotechnology. New England became the first industrialized region in the USA. The concentration of industries in this region is addressed as one of the factors that led to the foundation of the MIT in 1861 as school of industrial science. The MIT became the first university of technology in the USA, which came to strengthen both the industrial tradition established in the region and the academic tradition already existing with the presence of Harvard University in Cambridge.

Over the last century, Massachusetts has gone through different economic cycles related to periods of technological developments leading three waves of change and revitalization of industrial processes in the USA. At the beginning of the twentieth century, Massachusetts was the home of traditional industries in the textile and apparel sectors, which declined during the 1930s and 1940s (Castells and Hall 1994).

The first wave took place during the postwar period, which is associated with a phase of political and military tension that followed the end of the WWII. During this period, technology advanced to support military and space programs. According to Castells and Hall (1994), Massachusetts concentrated in research and manufacturing "mainly in precision instruments, avionics, missiles, and electrical machinery." The technology shift in warfare traces back to the 1930s and during the WWII. Then, MIT had the oldest and most distinguished electrical engineering department in the USA and was open to conduct contract research with the government and industry. MIT's advanced laboratories led the formation of important companies working in military research and manufacturing in the region. Many of these companies spun off from Raytheon, a major American technology company specialized in defense and security which started also at MIT in 1922 (Castells and Hall 1994), and the MIT Lincoln Lab, which by the end of 1980, had spun 39 new companies (Lampe 1988). During this period several of MIT's faculty members that led research programs at MIT worked as science advisors for the US government. The same research programs spun-off several research and manufacturing companies in the area. The close relationships of this network helped to enhance the technological potential of MIT, which advanced its electronic research with the support of the government.

The second wave took place during the ICT industrial revolution, which was characterized by advancements in microelectronics brought by space programs in the early 1960s. These advancements created fast-changing industries, which evolved from minicomputers to networks of computers, software, artificial intelligence, and telecommunication technologies. A loss of manufacturing jobs between 1967 and 1975 contributed to a reindustrialization process in the region. The technology-related businesses that emerged in the 1950s were affected because of the recession of military spending on research. A shift of technological paradigm was decisive leading the state's recovery from this downturn spiral. The establishment of Digital Equipment Corporation (DEC) in 1957 by MIT alumni was crucial in this shift. DEC was the first of many companies specialized in computer manufacturing that laid the grounds for an entire new industry that grew out of new knowledge and played a major role in the economic recovery of the state during the 1970s known as the Massachusetts Miracle. These new companies began locating in vacant factories along Highway 128 during the period of 1975–1980, giving shape to the high-tech complex known as Boston's Route 128. The employment in the state grew up because of the jobs generated by the fast technological changes of the ICT industrial revolution. Many of these new companies traced their origins also to research projects linked to MIT and/or other R&D firms (Lampe 1988).

The third wave has taken place during the Digital and Information Age that began with the invention of the WWW. In the region, the knowledge coming from the many universities and research institutes advanced technology in artificial intelligence, biotechnology, novel materials, and medical equipment. This wave can be seen as an evolutionary process of steady economy growth with an expansion of the industrial focus toward the creation of a biotechnology cluster. This process can be associated with the knowledge-driven reindustrialization, resulting from the global orientation of nations to use knowledge in strengthening their economies. In Massachusetts, it traces back to an economic development initiative introduced by the US Congress in 1980: the Bayh-Dole Act. This act provided legal basis and economic incentives for universities, giving them the right to own patents, grant licenses, and collect royalties arising from their federally sponsored research (Nelsen 2005). For research universities like a MIT, this led to an emphasis on licensing. In 1986, MIT reorganized its "Patent, Copyright, and Licensing Office" into a "Technology Licensing Office – TLO." The new office hired people with strong technical and business backgrounds and put emphasis on marketing and licensing of inventions while outsourcing the patent prosecution to law firms. As a result, the participation of faculty in patenting and licensing increased dramatically, i.e., about 100 licenses per year between 1986 and 2000. In 1987, the MIT TLO runs a policy experiment that allowed MIT to grant exclusive licenses to companies in which faculty members owned equity and to accept equity from licensed start-up companies as a form of royalty (Nelsen 2005). Since then, MIT has started more than 350 companies, which were formed to exploit MIT intellectual property in the fields of pharmaceutics, superconductors, batteries, Internet distribution, weather forecasting, and clean energy, among others. Informally, this office's role in starting companies is larger by encouraging the formation and growth of the start-up companies such as introducing them to investors and companies in raising capital.

Largely, the growth of formal technology transfer at MIT has had a significant contribution to the Massachusetts' biotechnology cluster, since most of the companies in the biotechnology sector have started as small, entrepreneurial companies within the past 15 years and a great deal of them were formed around MIT licenses (Nelsen 2005). The MIT campus has been the home of institutions and activities that have been a key in a complex process. It can be best described as an evolving knowledge-base and industrial renewal across different technological paradigms. Today, a trend toward energy research is perceived as a major theme among MIT's faculty and researches and can be the next technological paradigm spun out off campus.

10.4.2 MIT Sensing and Seizing the Opportunities While Orchestrating Its Campus as Strategic Resource

This section described two major CRE practices that facilitated the evolution of MIT into an entrepreneurial university. It describes – through shifts in the CRE practice – how the MIT has sensed and seized the opportunities given by its dynamic context and used the campus as a strategic resource supporting its evolving path.

Land Acquisition Strategy: From CRE Problem Solver to City Partner
In using their resources efficiently, most universities locate where they own a property. In the current competitive context, it matters where these properties are since some companies increasingly want to locate close to universities.

For MIT locating in Cambridge, it was an emergent accommodation decision that turned out to be a positive strategy for the institute and for Cambridge. The MIT was accommodated in Boston for more than 70 years since its foundation. The growing number of students in the beginning of the twentieth century created the need for expansion of the Institute's physical plant. In 1911, MIT acquired an 18-hectare plot (46 acres) located in East Cambridge that was surrounded by industrial districts. Soon, some of them became available and were acquired by the MIT (Fig. 10.3). In 1912, the Institute arranged the purchasing of additional land to the west of Massachusetts Avenue anticipating the MIT's future growth. This zone developed as the supporting environment for the academic life, providing housing, sports, and cultural-related functions. In the years that followed, MIT purchased additional land for "either immediate academic use or for investment use on an interim basis, awaiting the need for academic purposes" (Simha 2001). With this early intervention, MIT began a long-term acquisition plan for its land resources that secured its future growth many years in advance and determined its synergetic relationship with the city.

Today, MIT owns 104 hectares in Cambridge from which 68 hectares are tax-exempt. The use of the land resources by tax-exempt institutions such as the MIT has been a concern in Cambridge because taxes are the city's main source of income. Since 1928 MIT signed different agreements with the city of Cambridge to payments in lieu of taxes for a period of 20 years. Over the years, these friendly agreements, which are not a legal requirement, have sustained a good relationship between the institute and the city since they both have benefited from it. The university has secured its future growth many years in advance while enjoying a unique tax-exempt position. The city has received payments for land removed from the tax rolls and benefited of the presence of students, employees, and the business tenants

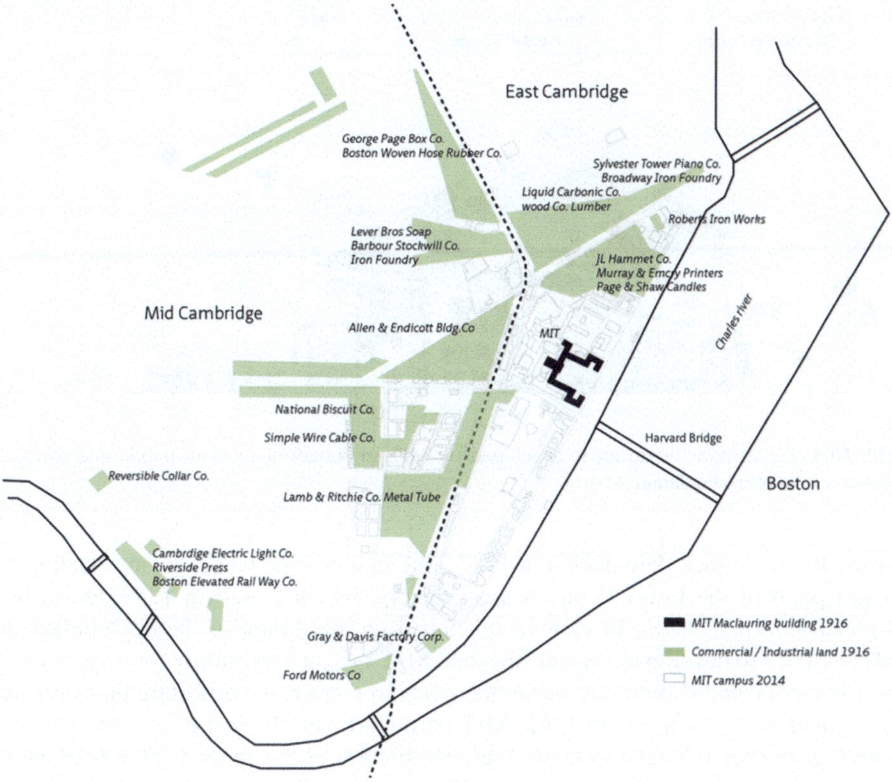

Fig. 10.3 Cambridgeport Land Use in 1916. Data base map: MIT Museum archives (Curvelo Magdaniel 2016)

of the MIT's commercial property. The latter are mainly large companies and research institutions that contribute to the local businesses in Cambridge by generating revenues for the city and attracting more companies to establish in the surrounding areas. Overall, the MIT's land acquisition strategy has been a good example of acting upon an emergent opportunity, which has mutually benefitted the MIT and the city of Cambridge.

Urban Area Development: From City Partner to Innovation Ecosystem Planner

MIT collaborated with public and private sectors in the development of three major urban areas surrounding the academic property: Technology Square, Kendall Square and University Park @MIT (Fig. 10.4). These areas have accommodated the changing activities resulted from the evolving knowledge base and industrial renewal across different technological paradigms in the region.

Technology Square

In the late 1950s, the city of Cambridge pursued an urban renewal project in East Cambridge. The area intended for development comprised residential land, known

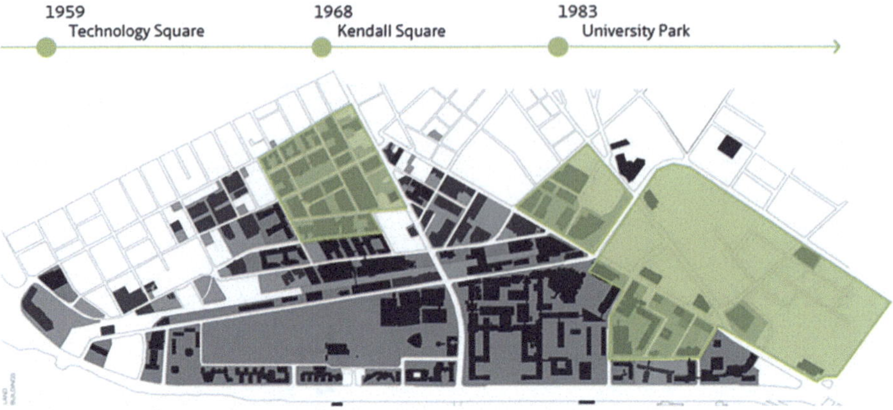

Fig. 10.4 Location of urban areas developed by MIT in collaboration with public and private partners (Curvelo Magdaniel 2016)

as the Roger's block demolished in 1957, and an industrial land, accommodating a major plant of the Lever Brothers soap factory, which closed in 1959 due to its relocation to New York. In view of this situation, the Major of the city contacted MIT in 1959 to develop the vacant site, and MIT saw an opportunity for investment. In 1960, MIT and Cambridge began the plans to convert the 6-hectare plot into an office and R&D complex. In 1962, MIT partnered with the real estate firm Cabot, Cabot & Forbes (CC&F) to begin the construction of the project. This marked a precedent, since it was the first time an educational institution worked together with a private firm to develop a business environment.

By 1967, the first four buildings of the complex were ready, and Polaroid set up its headquarters in Technology Square as well as others such as IBM, government agencies, and MIT's research groups. At the beginning of the 1970s, MIT sold its interest in Technology Square to CC&F but continued renting space for special research projects. During the 1980s Draper Laboratories – which became an independent research institute from MIT in 1973 – begins the construction of a new building in Technology Square. In 2001, MIT purchased the entire complex with the intention to maintain it as a tax-paying commercial property. This MIT's decision has benefited the city, which still receiving real estate tax revenues from this area and at the same time has reached its economic development goal of converting a former depressed industrial area into a R&D complex. Today, Technology Square is a mixed-use built environment that accommodates several offices, biotechnology lab, and street-level retail.

Kendall Square

In the 1960s, NASA funded research programs in Cambridge involving Harvard University and MIT. In the period 1964–1966, the Boston Redevelopment Authority

(BRA) oversaw the economic benefits for the region from NASA's presence and developed an urban renewal proposal for Kendall Square to clear its old industrial use. The proposal was presented to MIT because financing this urban renewal project would require MIT's cooperation and commitment to the City of Cambridge (Simha 2001). After MIT agreed to provide credits for the city, Cambridge invited NASA to locate its center here.

In 1968 the BRA, the Cambridge Redevelopment Authority (CRA) and NASA began the renovation works of the 12-hectare plot. The clearing of the old industrial uses took place between 1967 and 1975 during a period of economic recession. Within that period and after changes in the federal administration in 1969, NASA announced the termination of its Cambridge's activities. The new building erected for NASA's research center became vacant, but soon the CRA found a new occupant: the federal Department of Transportation (DOT) research center. In 1971 the DOT released 4.5 hectares of land that they considered a surplus. The CRA considered the option to sell the site to the MIT for academic purposes, but the Institute was interested in maintaining its academic activities compact. Instead, they proposed to convert the site into a housing neighborhood.

In 1974, the MIT and the East Cambridge Planning Team proposed a plan for a lively 24/7 neighborhood of mixed use including housing. This plan was based on 1964's study commissioned to Kevin Lynch, professor at the MIT School of Architecture and Planning at that time. In 1975, the CRA proceeded with the plan with changes allowing housing but not as a required function. In 1977 the zoning plan is approved proposing for first time mixed development zones for different land uses on a common site.

In 1979, Boston Properties were selected as a developer for the project. During the early 1980s, the proposal began development. Increasing traffic congestion issues triggered the planning of the new Kendall MBTA Red Line in 1983. In 1986, MIT participated in the design of the Kendall/MIT subway station, which construction in the 1990s would improve the connectivity of Cambridge in the region. As Boston Properties continued the mixed industrial and commercial development, the results have been criticized by the MIT for being monolithic and lacking sufficient services around the campus.

In 1999, an MIT alumnus established the Cambridge Innovation Centre (CIC) in Kendall Square. This became one of the earliest co-working spaces for start-up companies, offering affordable and flexible real estate for young entrepreneurs in the area, by renting space floor from MIT's owned building. By 2014, the CIC rented from MIT half of the space available in the same building to provide office space for over 500 companies, from which nearly 450 are start-ups. During the 2000s, several R&D companies, research institutions, and venture capital firms have located in Kendall Square. Most of them not only conduct businesses or research in the biotech and pharma sectors but also in IT and data and more recently in energy fields.

In 2011 the City of Cambridge released a planning study for Kendall Square, which area included 10-hectare parcel of MIT academic property. The same year the Institute filed a rezoning petition for this area. Herein, the MIT community raised its concern about the need for a long-term planning that considers the preservation of academic land resources and social inclusion that can be hindered with the commercial development emerging in Kendall Square area. Hence, a design committee for MIT's Kendall Square Initiative was established. This committee, formed by faculty from the MIT School of Architecture and Planning and the MIT community, is a form of participatory planning and design to ensure high quality of the built environment and alignment with the current planning and design principles of the MIT campus. The first outcome of this initiative is the MIT Gateway to Kendall Square Zoning petition, which was approved by the Cambridge City Council in 2013. The MIT's vision of mixed use neighborhood for Kendall Square persists from the 1970s up to date.

Kendall Square continues under development, and it is facing a spatial and functional transformation. It has become denser resembling the image of a financial and business district rather than a university environment. The strong presence of large corporations such as Google, Microsoft, and Novartis is dominant in the landscape. The public space is still poor in some areas, and the existing shops and restaurants and the new residential development are getting expensive for the students' and Cambridge's residents.

University Park at MIT

University Park at MIT is a mixed-use development of commercial, private laboratory, and incubator and residential functions, located in the parcels once occupied by the Simplex Wire & Cable Company. This company was a manufacturer of wire and cable for telephones established in Cambridge since 1888. In 1969, this company is sold to a company in New York that moved the operations to Maine. The property was placed in the market, and after the success experienced with Technology Square, MIT saw the potential of transforming the industrial district into a housing and commercial development. Between 1970 and 1971, MIT acquired the property.

MIT conducted a study aimed to identify the site's needs, considering the interests of the Cambridge's community on housing development. This led to a complex process of negotiations with the city and the community before the plan was completed. In 1983, MIT selected Forest City Enterprises (FCE) as developers for the site. In 1985, the City Council appoints a Planning Committee involving representatives of the MIT, FCE, and Cambridge's community. In 1987, these parties completed a master plan, which was approved by the City Council in 1989 and changed due to rezoning in 1992. The same year the development of the area began.

University Park at MIT is an example of real estate development in which the MIT established a long-term relationship with the community, because of the social component of housing development.

10.5 Discussion

The role of MIT's CRE decision-makers in supporting the universities evolutionary role has been illustrated above. Nevertheless, this study identified challenging issues in the CRE practice that are considered a risk for the MIT's sustainable competitive advantage.

Since the late 1990s, there has been a change in focus in land acquisition policies at MIT. The allocation of campus' land resources and area development efforts to commercial uses has been raising a conflict between specific stakeholders, whose perspectives on the campus supporting both the traditional and entrepreneurial roles of the university are incompatible.

MIT leaders and CRE managers perceive the commercial development of urban areas around Kendall Square as an opportunity to generate income that will sustain the Institute's mission while keeping an entrepreneurial environment around campus. This can be considered a short-term competitive position. For instance, during the development of MIT's Kendall Square initiative, the Institute abandoned its commitment to reserve the land south of Main Street for academic purposes. As a result, the 10-hectare parcel of MIT academic property will be converted into a mixed-use development including new housing, retail, lab, and commercial space. Similarly, academic land reserves have been leased to private firms for long terms, closing off MIT's academic expansion in the North Campus.

These initiatives have been supported by the City of Cambridge since it will benefit because of the revenues coming from property taxes while succeeding in their ambition to create an attractive "place to live, work, and do business" as part of their economic development strategy. A collaborative model of mutual understanding has strengthened the relationship between MIT and the City of Cambridge for decades. Nevertheless, the strategic nature of such relationship is increasingly built on financial ties, in which the lack of long-term planning can result in uncontrolled development that can be followed by political action. The recent involvement of the MIT community in the Kendal Square project is an initial MIT political stand to ensure the long-term (social and financial) sustainability of this area in line with the future growth and expansion of the Institute as well as the preferences of its main users (i.e., academic staff and students).

Academic leaders and influential members of the MIT community have perceived the emphasis on commercial area development as a threat for the nstitute's future in accommodating academic growth and fulfilling its academic mission. For this group, ensuring the institute's growth of academic space is as important as generating income to sustain its mission and focusing on short-term financial returns can be at the expense of long-term welfare of MIT.

The implications of this strategy for the MIT's financial sustainability can be roughly drawn in possible scenarios. If the MIT schools would need space that has been allocated for commercial use, they would have to pay for it at the high market price. Eventually, if schools have financial trouble in accommodating growing education or research programs, they will have to get the MIT's financial support. This can result in formal buying decisions in getting back those properties to academic use, which will decapitalize the Institute's endowment. Such decisions will raise political disagreements with the city that will suffer in the moment MIT decides to take such land off the city's tax roll. If not buying, the schools will have to lease more properties from other commercial parties at high costs. Potentially, the Institute must subsidize the cost of the rent, or the schools will have to raise capital outside MIT to buy buildings or rent space at the cost of their research programs. In this case, MIT schools and departments will be threatened by their academic competitors because the sponsors of research programs are unlikely to spend their budgets on space rather than on actual research.

Overall, there is a need for a healthy debate between these stakeholders to bring balance and to avoid uncontrolled development, which in the end can inhibit the role of MIT campus enabling both, research and education as well as other economic activities. The poor communication between these stakeholders and the cultural differences in their practices are not facilitating the required space for debate. CRE managers are aware of the risks on the long term but keep their position on investing MIT's capital according to the real estate opportunities while "learning by doing" on the process. The other stakeholders have manifested their opinions and reactions to this strategy through formal communication channels (e.g., the MIT Faculty Newsletter) in order to raise these concerns among members of the entire community (i.e., students, faculty, staff, alumni, parents, and more). Nevertheless, the periodic changes in administrations – both in MIT and the City of Cambridge – are major obstacles to have a continuous and healthy debate in overcoming this issue.

Simultaneously, the allocation of area development efforts to commercial uses is having an impact on campus life. Cambridge and Boston are exploiting the booming of the biotech and pharma cluster with urban development strategies aimed to sustain the presence of firms and research institutions in close proximity to the talent in academic institutions.

Nevertheless, the fostering of that mixed environment for "working, living, and doing business" is overlooking "studying" as part of that existing environment and more importantly as an essential activity in the city. Students and young people represent a considerable share of Cambridge's population. This trend is also visible in Kendall Square, where the MIT students' population is representative of the area's population. Nowadays the high concentration of firms and the intensification of commercial developments in Kendall Square are creating other problems. The area is becoming crowded and expensive. First, the dense concentration of users in the area generates traffic congestion with an environmental impact, which is not yet solved by the transit-oriented development. And second, the high rental prices of housing, office, and retail space are increasing the costs of living in the area, which can become unaffordable for the young community of entrepreneurs and students.

This situation is optimal for commercial real estate developers, who see physical proximity as an opportunity to boost their profitability. However, commercial real estate development needs to be controlled, especially around universities because many young people (e.g., students) cannot afford it. Largely, there is a need for more involvement from the municipality and representative of the communities to create a balanced mixed-use development, which considers the income difference among individuals who are an essential part of their promoted "entrepreneurial ecosystem." Investments in affordable housing, public transportation, and public space are crucial to enable a healthy environment for all the involved communities in Kendall Square area.

10.6 Conclusion

This chapter illustrated in an anecdotal way how the MIT – a leading entrepreneurial university – has used its campus as strategic resource to react timely and adequately to the dynamic context of the KBE. First, it provided an understanding of the coevolving path the MIT followed influenced by socioeconomic, technological, and institutional trajectories and how this path shaped its current position in the competitive higher education and research context. Second, it described with shifts in the CRE practice how campus development was used as a strategic resource to support the MIT's evolutionary path.

Similarly, this chapter identified challenges in the CRE practice that can threat the MIT's competitive position. It describes recent developments in such practice that can lead to possible scenarios and ways to address this situation. Largely, these remarks are important to understand that the entrepreneurial role of the MIT has coevolved with socioeconomic and technological developments in context, and it is not to be forced. Therefore, the CRE strategies encouraging this evolution have to be orchestrated carefully, balancing short- and long-term positions to attain sustainable competitive advantage. Hence, it brings the attention to acknowledge CRE practice facilitating or threatening organizational competitive advantage. Herein, this chapter attempts to bring forward the field of CREM in strategic management, since CRE is still a relatively less obvious resource for many organizations when developing a competitive strategy.

References

Altbach, P. G. (2004). The costs and benefits of world-class universities. *Academe, 90*(1), 20.

Briggs, S. (2006). An exploratory study of the factors influencing undergraduate student choice: The case of higher education in Scotland. *Studies in Higher Education, 31*(6), 705–722. https://doi.org/10.1080/03075070601004333.

Castells, M., & Hall, P. (1994). *Technopoles of the world.* London: Routledge.

Cooke, P., & Leydesdorff, L. (2006). Regional development in the knowledge-based economy: The construction of advantage. *The Journal of Technology Transfer, 31*(1), 5–15.

Curvelo Magdaniel, F. T. J. (2016). *Technology campuses and cities. A study on the relation between innovation and the built environment at the urban area level.* (PhD Doctoral thesis), Delft University of Technology, Delft.
Deem, R. (2001). Globalisation, new managerialism, academic capitalism and entrepreneurialism in universities: Is the local dimension still important? *Comparative Education, 37*(1), 7–20.
Den Heijer, A. (2011). *Managing the university campus. Information to support real estate decisions*. Delft: Eburon.
Dewulf, G. P. R. M., Krumm, P., & De Jonge, H. (2000). *Successful corporate real estate strategies*. Nieuwegein: Arko.
Drucker, J., & Goldstein, H. (2007). Assessing the regional economic development impacts of universities: A review of current approaches. *International Regional Science Review, 30*(1), 20–46.
Etzkowitz, H. (2004). The evolution of the entrepreneurial university. *International Journal of Technology and Globalisation, 1*(1), 64–77.
Etzkowitz, H. (2008). *The triple helix: University-industry-government innovation in action*. New York: Routledge.
Flyvbjerg, B. (2006). Five misunderstandings about case-study research. *Qualitative Inquiry, 12*(2), 219–245.
Headrick, D. R. (2009). *Technology: A world history*. New York: Oxford University Press.
Helfat, C. E., Finkelstein, S., Mitchell, W., Peteraf, M., Singh, H., Teece, D., & Winter, S. G. (2009). *Dynamic capabilities: Understanding strategic change in organizations*. New York: Wiley.
Jessop, B. (2017). Varieties of academic capitalism and entrepreneurial universities. *Higher Education, 73*(6), 853–870. https://doi.org/10.1007/s10734-017-0120-6.
Joroff, M. (1993). *Strategic management of the fifth resource : Corporate real estate*. Norcross: Industrial Development Research Foundation.
Kauppinen, I. (2012). Towards transnational academic capitalism. *Higher Education, 64*(4), 543–556. https://doi.org/10.1007/s10734-012-9511-x.
Lampe, D. (1988). *The Massachusetts miracle: High technology and economic revitalization*. Cambridge, MA: The MIT Press.
Marginson, S., & Considine, M. (2000). *The enterprise university*. Cambridge: Cambridge University Press.
Nelsen, L. L. (2005). The role of research institutions in the formation of the biotech cluster in Massachusetts: The MIT experience. *Journal of Commercial Biotechnology, 11*(4), 330–336.
Nourse, H. O., & Roulac, S. E. (1993). Linking real estate decision to corporate strategy. *The journal of real estate research, 8*(4), 475–494.
OECD. (1996). The knowledge-based economy. Retrieved from Paris:
OECD. (2013). Science and Technology. Researchers. Retrieved October 20, 2015 https://www.google.com/publicdata/explore?ds=ltjib1m1uf3pf_&ctype=l&met_y=research_t1
Porter, M. (1990). *The competitive advantage of nations*. New York: Free Press.
Rumelt, R. P. (1984). Towards a strategic theory of the firm. Competitive strategic management. *Competitive Strategic Management*, 556–570.
Rumelt, R. P., Schendel, D., & Teece, D. J. (1994). *Fundamental issues in strategy: A research agenda*. Boston: Harvard Business Press.
Sam, C., & Van Der Sijde, P. (2014). Understanding the concept of the entrepreneurial university from the perspective of higher education models. *Higher Education, 68*(6), 891–908.
Schulze-Cleven, T., & Olson, J. R. (2017). Worlds of higher education transformed: Toward varieties of academic capitalism. *Higher Education, 73*(6), 813–831. https://doi.org/10.1007/s10734-017-0123-3.
Schulze-Cleven, T., Reitz, T., Maesse, J., & Angermuller, J. (2017). The new political economy of higher education: Between distributional conflicts and discursive stratification. *Higher Education, 73*(6), 795–812. https://doi.org/10.1007/s10734-017-0114-4.

Schumpeter, J. A. (1934). The theory of economic development: An inquiry into profits, capital, credit, interest, and the business cycle. In *Transaction publishers* (Vol. 55).

Shuen, A., & Sieber, S. (2009). Orchestrating the new dynamic capabilities. *IESE Insights, Fourth Quarter, 3*, 58–65.

Simha, O. R. (2001). *MIT campus planning. An annotated chronology*. Cambridge: Massachusetts Institute of Technology.

Simha, O. R. (2005). The economic impact of eight research universities on the Boston region. *Tertiary Education and Management, 11*(3), 269–278. https://doi.org/10.1080/13583883.2005.9967151.

Simmie, J. (2005). Innovation and space: A critical review of the literature. *Regional Studies, 39*(6), 789–804.

Srivastava, R. K., Shervani, T. A., & Fahey, L. (1998). Market-based assets and shareholder value: A framework for analysis. *The Journal of Marketing, 62*(1), 2–18.

Teece, D. J. (1984). Economic analysis and strategic management. *California Management Review, 26*(3), 87–110.

Teece, D. J. (2007). Explicating dynamic capabilities: The nature and microfoundations of (sustainable) enterprise performance. *Strategic management journal,* 28(13), 1319–1350.

Teece, D. J., Pisano, G., & Shuen, A. (1997). Dynamic capabilities and strategic management. *Strategic Management Journal, 18*(7), 509–533.

Too, L., Harvey, M., & Too, E. (2010). Globalisation and corporate real estate strategies. *Journal of Corporate Real Estate, 12*(4), 234–248. https://doi.org/10.1108/14630011011094676.

Van Winden, W. (2008). Urban governance in the knowledge-based economy: Challenges for different city types. *Innovations, 10*(2–3), 197–210.

Vorley, T., & Nelles, J. (2008). Institutional development of and beyond the third mission. *Higher Education Management and Policy, 20*(3), 119.

Chapter 11
Increasing the Economic Sustainability of the Company JSC "Novoazovskoe"

Victoria Kopeikina, Paula Odete Fernandes, and Olga Kosenchuk

11.1 Introduction

The problem of economic sustainability of enterprises is substantially worrying problem. The situation of agricultural enterprises in current conditions is quite unsteady, highly risky and accounts for low profitability. These and other factors lead to rethink the main principles of economic sustainability and activity of the enterprise. As economic sustainability is equivalent to balanced economic resources that ensure a stable generating of profit and normal conditions for expanded production, persistent growth in long-term prospective took into account main external factors.

The relevance of research about the increase of economic sustainability of an agricultural enterprise is conditioned by the fact that development and economically efficient activity of the enterprise are in direct relation to the ability of implementing the most modern agricultural technologies. Nowadays, the leading enterprises in this branch, both domestic and international, realized the importance of implementing the latest scientific achievements in production. By doing so, these companies optimize the expenditures on resources, finances and time. This leads to the opportunities for intensification of production and, consequently, improving product quality, generating additional profit and strengthening the company's position in the market.

V. Kopeikina · O. Kosenchuk
Omsk State Agrarian University, Omsk, Russia

P. O. Fernandes (✉)
UNIAG, Applied Management Research Unit; NECE, Research Unit in Business Sciences (UBI), Polytechnic Institute of Bragança (IPB), Campus de Santa Apolónia, Bragança, Portugal
e-mail: pof@ipb.pt

M. Peris-Ortiz et al. (eds.), *Knowledge, Innovation and Sustainable Development in Organizations*, Innovation, Technology, and Knowledge Management,
https://doi.org/10.1007/978-3-319-74881-8_11

In this regard, the main objective of this chapter is to search for theoretical statements and validation measures of economic sustainability regarding JSC "Novoazovskoe".

Tasks of the chapter are:

1. Consider the main problem and ways of increasing economic sustainability of enterprises generally and the agricultural enterprise particularly.
2. Analyse the activity of JSC "Novoazovskoe".
3. Analyse the ways of increasing the economic sustainability of enterprises.
4. Justify measures of increasing the economic sustainability of JSC "Novoazovskoe" by implementing a modern system of haylage harvesting.

JSC "Novoazovskoe" of Azovsky German National district of Omsk region of Russian Federation it was the enterprise under study.

The research objective is the economic sustainability of JSC "Novoazovskoe". The subject of research is the implementation of a modern system of haylage harvesting as the way of increasing economic sustainability of enterprises.

In the first section, various definitions of "economic sustainability" and "economic solidity" will be considered, and specific features of increasing economic sustainability of agricultural enterprise will be named. All the analyses approaches will be used in the second section, and it will be analysed the legal status of the enterprises as well as its managerial structure, production activity, and financial results of activity covering the years from 2013 to 2015. The analysis of potential ways of increasing economic sustainability will also be implemented. In the fourth part, it will be put forward the ways of increasing economic sustainability of JSC "Novoazovskoe" by using the modern haylage harvesting system. Concurrently the total financial and production results obtained after the implementation of the new haylage harvesting system will be calculated.

11.2 Theoretical Base for a Sustainability of a Company

11.2.1 Meanings of Economic Sustainability and Competitiveness of a Company

Nowadays, due to the impact of external environment on enterprises, the latter turn into more complicated systems. The process of integration of enterprises by mergers and acquisitions is undergoing. This is conducted either voluntarily or forcibly. All these meet the new requirements of the enterprises managing methods in current conditions. Because of that one of the most important issues for any enterprise is to ensure its own economic sustainability in short-term and long-term prospective (Anisimov 2006).

Firstly, the term "economic sustainability" appeared because of the consideration of the problem of scarcity of recourses, which became the result of global energy crisis of 1973 and 1979. Then, this flow of economic thought became a separate discipline called "ecoestate", which means the economic sustainability of a country. It started to consider the issues of sustainable economic development at the level of a country and a region.

However, recently it became increasingly clear that economic sustainability development of a country and a region is achieved only with economic sustainability of all structural elements such as industries, enterprises and organizations. Economic sustainability of each individual enterprise allows all economic system of a country not only to save the potential, but also to ensure its quality rise and entrance to international markets with new competitive goods (Laibert 2011). Nowadays there is a multiplicity of viewpoints of different authors regarding the concept "economic sustainability of an enterprise".

It is thought that economic sustainability means not only the keeping of positive trends of all related indicators, but also includes development, which appears in economic growth, that is a trend of positive changes of aggregate indicators of development for a determined period of time (usually for a year). For the economic growth features, both common and particular indicators are used. The most interesting and important indicator among all is financial solidity.

The comparison between the concepts of "economic sustainability" and "financial solidity" shows us that most authors strengthen these definitions. The rationale for this is the statement that the enterprise's position in the market firstly depends on the presence and directions of financial resource usage. Estimation of financial position of an enterprise can help to determine the "bottlenecks" in entrepreneurial activity and to find the solution for avoidance of adverse trends of the company's development. Thus, financial solidity permits objectively to estimate the tactic of management (Lakshina and Chekmareva 2009).

Sustainable business of an enterprise depends on the internal possibilities to effectively use all the available economic resources. Under the economic resources are referred all natural, human and man-made resources, which are used for the production of goods and services.

Under economic resources are referred all kinds of resources used in the process of goods and service production. Basically, it is the goods, which are used for other goods production (Bulatov 2007).

Thus, the definition of economic sustainability can be formulated as the balanced state of economic resources, which ensure stable profitability and normal conditions for sustainable economic growth in long-term perspective taking into account the most important internal and external factors.

11.2.2 Types of Economic Sustainability and Factors of Sustainable Development of a Company

Many definitions of "economic sustainability" in relation to enterprises have the characteristics (parameters) of sustainability. That is, on the level of companies, it is possible to distinguish various components of sustainability: financial, technological, organizational, commercial and others. Consider some of them more in detail (Golovko 2013).

Production and technical sustainability of an enterprise is determined by its solidity of production cycle, coordination of resource and material provision.

Commercial sustainability is determined by the level of economic activity, reliability of economic ties, competitive potential of a company and its share in the market.

Organizational sustainability means the stability of internal organizational structure, relation and efficiency of ties between departments and offices of an enterprise and efficiency of their common operations.

Innovation sustainability is characterized by the ability of an enterprise to implement new technologies and methods of production organization, release new kinds of production, perform new kinds of work, provide new kinds of services and be ready for innovations and modifications.

Financial sustainability is characterized by the state of financial resources of an enterprise, which ensures their efficient use for an uninterrupted process of production and realization of goods.

Social sustainability means the involvement of all collectives of an enterprise into the social process, assistance for the growth of welfare of society, insurance of development and level of social welfare of employees.

All parameters of economic sustainability are related and interacted. The level of development of each parameter impacts on total economic sustainability of an enterprise. Moreover, total economic sustainability of the enterprise makes the impact on the sustainability of its industry and sustainability of the location of a region. Industrial and regional sustainability forms the common economic sustainability of a country.

Thus, sustainability of economic system can be determined as the state of a system, on which related parameters (financial, production, organizational and others) are, beforehand, assigned limits of sustainability and, simultaneously, are available to harmonious development and perfection, under any changes of external environment.

In the context of sustainable development of enterprise, it is necessary to identify the factors, which impact on its provision, as the following (Koryakov 2012):

1. Factors, which affect the social component of sustainable development of an enterprise
2. Factors, which affect the ecological component of sustainable development of an enterprise
3. Factors, which affect the economic component of sustainable development of an enterprise

According to some researchers, the approach to the analysis of factors of sustainable growth requires a systematic classification of factors based on market categories. As such, these factors can be differentiated into three groups (Koryakov 2012):

- Supply factors
- Demand factors
- Distribution and redistribution factors

The supply factors determine the potential for economic growth.

It should be noted that supply and demand factors are interconnected. For instance, unemployment slows the rate of capital accumulation, reduces the flow of investments and slows the expansion of production. Conversely, low rate of investments may be the main cause of unemployment.

By exploring the problem of typology of factors of sustainable development of a company, it is necessary to turn to the economic theory, which traditionally distinguishes three groups of production factors: land, labour and capital (Marks and Engels 1984).

Based on the research of works related to sustainable development, we found it necessary to add to the list of factors the following ones: innovations and management, information, ecology and politics (Koryakov 2012).

11.2.3 Features of Increasing the Sustainability of a Company in Agribusiness Industry

There are three basic components of sustainable development characterized by the following factors:

- "Social" development – labour
- "Ecological" development – natural factors and ecology
- "Economic" development – capital, innovations and management, information and political lobby (Koryakov 2012)

The key to survival and the basis for solid state of an enterprise are their sustainability, which is influenced by different factors: position of the company in financial market, production of quality demanded products, potential of enterprise to business cooperation, degree of dependence on external creditors and investors, presence of insolvent creditors, effectiveness of economic and financial transactions, etc. (Folomjev 1995).

Factors of constancy of economic development of the enterprise also represent an array of threats and opportunities of external and internal environment. Important factors are political situation, rational use of natural resource potential of the enterprise, results of market reforms of ownership relations, improvement of living conditions of the population and preservation of ecological safety of the territory of an enterprise. With such a variety of factors, it is necessary to differentiate the economic sustainability of the enterprises by types.

Factors of sustainable development of the enterprise should be explored in such aggregated areas:

- Global conditions (state of the world economy, globalization of the economy, external threats, participation in cross-border and global cooperation, regional integration)
- Material and physical capital (quantity and quality of land, size, structure, status, effectiveness of fixed assets, etc.)

- Human capital (number and structure of population, structure of employment, cultural and professional level of population)
- Financial resources of enterprise
- Perfection of market relations (level of concentration, market infrastructure, level of privatization, level of state regulation)
- Noneconomic factors (political, social and cultural environment) (Koryakov 2012)

The process of sustainable economic development, as well as the level of competitiveness of the national economy, depends on such groups of macro-level factors:

- Level of provision of the economy with natural resources, labour, production, information and other resources.
- Level of implementation's efficiency by the government's demographic, investment, innovation and regulatory policy. Due to that, with a country that is optimizing the demographic structure of the population and forming the regulatory environment that encourages the business to function effectively, other countries and businesses can be investing in promising sectors of the national economy.
- Opportunities for business owners to realize the competitive advantages of the existing resource potential and the regulatory environment with the aim to implement this or that type of intense economic reproduction and achieve, on this basis, a high level of efficiency and, eventually, sustainable economic development.

Sustainable development, as well as the competitiveness of a country, is the result of permanent interaction of all subjects of economic relationship in the country. Consequently, the research of economic categories should consider the analysis of all factors, which influence the formation of the conditions for sustainable economic development of a country in each stage of extended production and in all levels of competition, namely, enterprise, industry, form of economic activity and the entire economy (Koryakov 2012).

Considering the definition of "sustainability" according to agriculture, it can be characterized as the ability to resist against negative impact of various production factors. Sustainability of agricultural production influences the set of various factors, which can be classified into external and internal (Kuzmenko and Gritcenko 2016).

Sustainable agriculture integrates three main goals: environmental health, economic profitability and social and economic equity.

There are specific strategies for realizing these broad themes or goals. The strategies are grouped according to three separate though related areas of concern: farming and natural resources, plant and animal production practices and economic, social and political context. They represent a range of potential ideas for individuals committed to interpreting the vision of sustainable agriculture within their own circumstances.

Sustainable production practices involve a variety of approaches. Specific strategies must take into account topography, soil characteristics, climate, pests, local availability of inputs and the individual grower's goals. Despite the site-specific and individual nature of sustainable agriculture, several general principles can be applied to help growers select appropriate management practices (Feenstra et al. 2016):

- Selection of species and varieties that are well suited to the site and to conditions in the farm
- Diversification of crops (including livestock), harvesting and cultural practices to enhance the biological and economic stability of the farm
- Management of the soil to enhance and protect soil quality
- Efficient and humane use of inputs
- Consideration of farmers' goals and lifestyle choices

Technical factors, which include improvement of existing and creation of new machineries and implementation of resource-saving technologies of agricultural crop cultivation, help to increase agricultural production sustainability. In Sect. 11.3 the weaknesses of investigated enterprise will be identified, which help to inspect the most prospective direction of activity aimed at improving economic sustainability of the enterprise (Kuzmenko and Gritcenko 2016).

11.3 Organizational and Economic Evaluation of JSC "Novoazovskoe"

11.3.1 Legal Status and Management System of JSC "Novoazovskoe"

For further research of the enterprise, it is necessary to conduct the analysis of organizational structure on the company. The main characteristics of the structure's quality of any economic system are the equilibrium and proportionality of its parts' interconnections (subdivisions and employees). The organizational structure of an enterprise is the ordered collection of solidity connected subsystems ensuring the functioning and development of the organization as an aggregate.

JSC "Novoazovskoe" is characterized by the linear-functional type of organizational structure. Linear-functional management structure ensures such a diversification of managerial work, in which the linear managerial elements provide overall leadership and coordination, and the functional ones consult and develop specific questions (Goldstein 2003).

As follows from the previously mentioned, the structure of JSC "Novoazovskoe" is linear with direct subordination, balanced in terms of number of services and departments, as well as a number of linear managers.

11.3.2 Characteristic of Activity of JSC "Novoazovskoe"

JSC "Novoazovskoe" is located in the territory of Azovsky German National District of Omsk Region. Azovsky District is in the south forest-steppe zone of Omsk region. The climate of Azovsky District is typically continental and formed by the cold arctic air masses from the North and from Kazakhstan. Common features of temperature regime are characterized by hard and long winter and short but hot summer. Common positive features are the abundance of light and heat during the vegetative period with positive temperatures, which accelerates vegetation of plants. The average monthly temperature of the warmest month is (July) 18 degrees and the coldest month (January), 19 degrees below zero. A stable snow cover forms at 6–12 of November. Annual precipitation is 300 mm. Prevailing winds from autumn to spring are the southwestern winds and in summer northwestern ones. Soil cover is represented mainly by ordinary chernozems (black soils) with a predominance of heavy mechanical composition, which is resistant to wind and water erosion (OmskPortal 2017).

To make the estimation of economic and business efficiency of the enterprise's activity, it was necessary to make the analysis of the following indicators: changing of balance sheet structure from year to year, changing in financial results from year to year, dynamics of changes in number and amount of current assets, dynamics of changes in square and quality of lands, dynamics of changes in average number of employees and production of various kinds of agricultural products.

Firstly, it needs to inspect the balance sheets and the income statements of JSC "Novoazovskoe" as the most important and reliable source of information. In addition, it needs to conduct the vertical analysis and horizontal analysis to get more information about the trends of enterprise development. The analysis of the balance's structure was made for 3 accounting years. Data is obtained from accounting statements and presented in Table 11.1 (year-to-year horizontal analysis). "Year-to-year" analysis means that all years will be compared in the sequence of ascending. Results are presented in Table 11.1.

For further analysis of business activity of an enterprise, it is necessary to analyse the income statement of JSC "Novoazovskoe" for the last 3 years. We need to make the vertical analysis and horizontal analysis to get more information about the trends of enterprise development. The analysis of the income statement's structure was made for 3 accounting years. Data obtained from accounting statements is presented in Table 11.2 (year-to-year horizontal analysis). All the items in income statement were compared with "total revenues", because in this case it can see the structure of revenues and expenditures and their influence on total income. In the section "Additionally", the provision for a single agricultural tax is indicated that is used in Russian Federation for agricultural companies and equal to 6% of "net profit".

The next step of analysis of business activity and state of an enterprise is to analyse its current assets. Firstly, it will be started with analysing the lands' composition and structure. For inspection of square and quality of lands, it is necessary to make the analysis of its structure and composition for the last 3 years. Secondly, it needs to determine the presence of the means of production (combine harvester, tractors, etc.) and its dynamic of change for the last 3 years. Because JSC "Novoazovskoe" produces milk, the number of animals employed also needs to be inspected.

For further analysis of an enterprise, it is also paramount to analyse the labour resources of JSC "Novoazovskoe". Inspection of statements with specified data reveals also crucial for the analysis of the composition and structure of revenues from sale to production.

Table 11.1 Year-to-year horizontal analysis of JSC "Novoazovskoe" balance sheets

Balance sheets in thousands of RUB	Year			2014 to 2013		2015 to 2014	
	2013	2014	2015	In amount	% of previous	In amount	% of previous
Assets							
Noncurrent assets							
Fixed assets	135,113	150,735	160,144	15,622	111.6%	9409	106.2%
Financial investments	0	0	0	0	0.00%	0	0.00%
Other assets	0	9115	15,874	9115	–	6759	174.2%
Total noncurrent assets	**135,113**	**159,850**	**176,018**	24,737	118.3%	16,168	110.1%
Current assets							
Inventory	125,688	130,491	147,155	4803	103.8%	16,664	112.8%
Accounts receivable	9673	7676	9838	−1997	79.4%	2162	128.2%
Cash and cash equivalents	49,069	50,544	35,449	1475	103.0%	−15,095	70.1%
Prepaid expenses and other current assets	0	0	0	–	–	–	–
Total current assets	**184,430**	**188,711**	**192,442**	4281	102.3%	3731	102.0%
Total assets	**319,543**	**348,561**	**368,460**	29,018	109.1%	19,899	105.7%
Liabilities and equity							
Equity							
Common stock	102	102	102	0	100.0%	0	100.0%
Revaluation of noncurrent assets	35,898	32,547	29,462	−3351	90.7%	−3085	90.5%
Reserved capital	14,020	18,955	21,924	4935	135.2%	2969	115.7%
Accumulated profit	268,614	296,531	313,143	27,917	110.4%	16,612	105.6%
Total equity	**318,634**	**348,135**	**364,631**	29,501	109.3%	16,496	104.7%
Long-term liabilities							
Long-term debt	0	0	0	0	0.00%	0	0.00%
Other long-term liabilities	0	0	0	0	0.00%	0	0.00%
Total long-term liabilities	**0**	**0**	**0**	0	0.00%	0	0.00%
Short-term liabilities							
Accounts payable	832	426	3829	−406	51.2%	3403	898.8%
Deferred revenue	77	0	0	−77	0.0%	0	0.0%
Short-term debt	0	0	0	0	0.0%	0	0.0%
Total short-term liabilities	**909**	**426**	**3829**	−483	46.9%	3403	898.8%
Total liabilities and stockholders' equity	**319,543**	**348,561**	**368,460**	29,018	109.1%	19,899	105.7%

Table 11.2 Year-to-year horizontal analysis of income statement of JSC "Novoazovskoe"

In thousands of RUB	Year			2014 to 2013		2015 to 2014	
Name of item	2013	2014	2015	In amount	% of previous	In amount	% of previous
Revenues							
Plant growing	30,896	56,624	38,823	25,728	183.3%	−17,801	68.6%
Animal breeding	187,212	140,732	151,935	−46,480	75.2%	11,203	108.0%
Other goods	6	0	0	−6	0.0%	0	–
Other services	661	638	757	−23	96.5%	119	118.7%
Total revenues	**218,775**	**197,994**	**191,515**	−20,781	90.5%	−6479	96.7%
Cost of revenues							
Plant growing	28,205	53,977	40,472	25,772	191.4%	−13,505	75.0%
Animal breeding	168,524	114,321	130,987	−54,203	67.8%	16,666	114.6%
Other goods	7	0	0	−7	0.0%	0	–
Other services	1053	1080	1175	27	102.6%	95	108.8%
Total cost of revenues	**197,789**	**169,378**	**172,634**	−28,411	85.6%	3256	101.9%
Gross profit							
Plant growing	2691	2647	−1649	−44	98.4%	−4296	−62.3%
Animal breeding	18,688	26,411	20,948	7723	141.3%	−5463	79.3%
Other goods	−1	0	0	1	–	0	–
Other services	−392	−442	−418	−50	–	24	–
Total gross profit	**20,986**	**28,616**	**18,881**	7630	136.4%	−9735	66.0%
Other items							
Interest income	682	2805	5236	2123	411.3%	2431	186.7%
Interest expense	0	0	0	0	0.0%	0	0.0%
Other incomes	15,684	9909	8320	−5775	63.2%	−1589	84.0%
Other expenses	3268	7672	11,575	4404	234.8%	3903	150.9%
Profit before income taxes	**34,084**	**33,658**	**20,862**	−426	98.8%	−12,796	62.0%
Other income (expense), net	1917	755	1075	−1162	–	320	–
Net profit	**32,167**	**32,903**	**19,787**	736	102.3%	−13,116	60.1%
Additionally							
Provision for single agricultural tax	1930	1974	1187	44	102.3%	−787	60.1%

For a more detailed analysis of an enterprise's efficiency, it is necessary to consider such indicators as production yield per one hectare and average milk yield per one cow. In Table 11.3 the calculation of an average milk yield and yield of cereals and vegetables from one hectare are put forth.

To complete the analysis and to evaluate the competitive advantages of JSC "Novoazovskoe", the SWOT analysis was performed. SWOT analysis is an acronym for strengths, weaknesses, opportunities and threats and is a structured planning method that evaluates those four elements of an organization, project or business venture. A SWOT analysis can be carried out for a company, product, place, industry or person. It involves specifying the objective of the business venture or project and identifying the internal and external factors that are favourable and unfavourable to achieve that objective (Osita et al. 2017). The results of implementation of this analysis are presented below.

- Strengths
  - Big production and managerial experience
  - Balanced system of milk production and feed harvesting
  - Presence of permanent and reliable suppliers of equipment and materials, as well as permanent buyers
  - Relatively close to the city of Omsk – less than 15 km – the major centre of production distribution
- Weakness
  - Low quality of harvested feeds because of the big loss of harvested feeds as the result of improper and old-style process of harvesting and storage
  - Unacceptable conditions of storage of haylage and its loss of more than half of nutrition value
  - Lack of modern equipment for feed harvesting, storage and delivery
- Opportunities
  - Ability to serve additional groups of consumers – farmers of Ural, Far East of Russia, as well as neighbouring countries: Kazakhstan, Uzbekistan, Tadzhikistan, Kyrgyzstan, Mongolia and China;
  - Opportunity to enter to other markets of feeds, for example haylage and silage.
- Threats
  - Change of climate and weather conditions, crop failure, low milk yield and loss of cattle
  - Strengthening of positions of local competitors with equal costs or investments of foreign producers with supermodern technologies, leading to decreasing the cost of production and increasing high quality
  - The lack of sufficient support and crop insurance from the state
  - Decline of purchasing prices by dealers
  - Increase of energy prices, fertilizers and feeds
  - Accelerated rate of inflation

The main threats are associated with the production of the main product, milk. A weakness lies in the low quality of harvested haylage, as well as the absence of certain types of agricultural machinery, necessary for a complete modern harvesting, transportation and storage of feed. In spite of balanced and proven technology of harvesting and feeding, the company remains competitive mainly due to the passive presence of foreign competitors in the market. However, the management's

Table 11.3 Production results of specified kind of productions in JSC "Novoazovskoe"

Parameter	Year			2014 to 2013		2015 to 2014		2015 to 2013	
	2013	2014	2015	In ha	In %	In ha	In %	In ha	In %
Total square of using lands (in hectares)									
Cereals total	5931	5815	5800	−116	98.0%	−15	99.7%	−131	97.8%
Spring grains	5581	5315	5100	−266	95.2%	−215	96.0%	−481	91.4%
Legumes grains	350	500	700	150	142.9%	200	140.0%	350	200.0%
Rapeseed	269	385	400	116	143.1%	15	103.9%	131	148.7%
Perennial grasses	1239	1317	1257	78	106.3%	−60	95.4%	18	101.5%
Annual grasses	1981	1914	1903	−67	96.6%	−11	99.4%	−78	96.1%
Corn for silage and green fodder	660	600	600	−60	90.9%	0	100.0%	−60	90.9%
Total harvest of plant production (in centners)									
Cereals total	139,930	119,290	103,395	−20,640	85.2%	−15,895	86.7%	−36,535	73.9%
Spring grains	133,539	112,420	93,647	−21,119	84.2%	−18,773	83.3%	−39,892	70.1%
Legume grains	6391	6870	9748	479	107.5%	2878	141.9%	3357	152.5%
Rapeseed	5670	8940	1371	3270	157.7%	−7569	15.3%	−4299	24.2%
Perennial grasses as a hay	11,214	3217	8153	−7997	28.7%	4936	253.4%	−3061	72.7%
Perennial grasses as a green mass	90,509	72,092	80,877	−18,417	79.7%	8785	112.2%	−9632	89.4%
Annual grasses	145,781	118,450	152,732	−27,331	81.3%	34,282	128.9%	6951	104.8%
Corn for silage and green fodder	96,988	61,842	170,546	−35,146	63.8%	108,704	275.8%	73,558	175.8%

Total yield of plant production (in centner/ha)									
Cereals total	23,6	20,5	17,8	–3	86.9%	–3	86.8%	–6	75.4%
Spring grains	23,9	21,2	18,4	–3	88.7%	–3	86.8%	–6	77.0%
Legume grains	18,3	13,7	13,9	–5	74.9%	0	101.5%	–4	76.0%
Rapeseed	21,1	23,2	3,4	2	110.0%	–20	14.7%	–18	16.1%
Perennial grasses as a hay	33,5	10,9	34,0	–23	32.5%	23	311.9%	1	101.5%
Perennial grasses as a green mass (non-applicable)	x	x	x	x	x	x	x	x	x
Annual grasses (non-applicable)	x	x	x	x	x	x	x	x	x
Corn for silage and green fodder	147,0	103,1	284,2	–44	70.1%	181	275.7%	137	193.3%
Total yield of milk production									
Number of milking cows, units	1023	1023	1023	0	100.0%	0	100.0%	0	100.0%
Milk yield, centners	53,354	55,812	59,382	2458	104.6%	3570	106.4%	6028	111.3%
Average yield of one cow per year, litres	5215	5456	5805	240	104.6%	349	106.4%	589	111.3%
Average yield of one cow per day, litres	14,3	14,9	15,9	0,7	104.6%	1,0	106.4%	1,6	111.3%

policy is focused on the creation of hi-tech and modern production, new markets and new products, constant updating of agricultural machinery, use of new varieties of crops and use of science-based technologies of their cultivation. These are all factors which, in consequence, turn the company into a strong competitor even for foreign manufacturers.

11.4 Increasing the Economic Sustainability of JSC "Novoazovskoe"

11.4.1 Description of Existing Harvesting System and Identifying the Reasons for Managerial Decisions in JSC "Novoazovskoe"

The SWOT analysis implemented and helped to detect the weakness of JSC "Novoazovskoe", among which are:

- Low quality of harvested feeds because of a big loss of harvested feeds as the result of an improper and old-style process of harvesting and storage
- Unacceptable conditions of storage of haylage and its loss of more than half of nutrition value
- Lack of modern equipment for feed harvesting, storage and delivery

!!As the result of research of organizational and economic state of an enterprise, the calculation of innovative activity indicators, as well as the state of plant growing and livestock production branches, together with specialists and top management of the company, possible development guidelines were discussed. The increase of quality of harvesting feeds was selected from all the priority trends. It will consider some of the propositions. For harvesting and storage of feeds, it is necessary to:

- Improve the quality of harvested haylage by equal distribution and drying of all harvested mass on the field to 50–55% of moisture.
- Purchase of a baler, helping to solve the problem by saving the quality of feeds during all storage term.
- Foreign substances exclusion in haylage, preventing the formation of pathogenic microorganisms negatively impacting on the health of animals (Agrovesti.net 2017).

It will consider all of the aforementioned measures in more detail. Negative results of the use of existing type of harvesting process are schematically presented in Fig. 11.1.

In monetary terms, the cost of repairs and idle time of machineries and employees during haylage harvesting process are presented in Table 11.4. Data was obtained from managerial statement of JSC "Novoazovskoe". As it can see from Table 11.4, accumulated result of repairs of machineries and idle time of machineries and

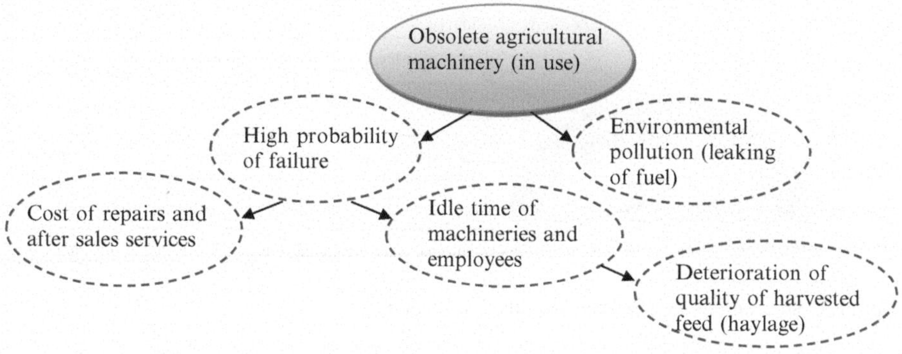

Fig. 11.1 Negative results of obsolete machineries use in haylage harvesting process

Table 11.4 Cost of repairs and idle time during haylage harvesting process

Indicator	2013	2014	2015
Machinery idle time, hours			
Tractor MTZ-82	8	15	10
Mower	0	15	8
Salary, rubles/hour			
Tractor-driver	96	118	129
Mower operator	96	118	129
Salary in idle time, rubles			
Tractor-driver	771	1763	1292
Mower operator	0	1763	1034
Cost of repairs, rubles			
Tractor MTZ-82	24,500	0	6100
Mower	0	42,300	13,800
Total, rubles			
Tractor MTZ-82	25,271	1763	7392
Mower	0	44,063	14,834
Total for harvest, rubles	**25,271**	**45,826**	**22,226**
Accumulated result	25,271	71,096	93,322

employees have a cost to the company of 93,322 rubles. Even though this does not represent crucial expenditures, nonetheless the company should spend money on it, interrupt the harvesting process and as a result, obtain improperly harvested haylage.

At the present time, JSC "Novoazovskoe" is using the following scheme of haylage harvesting:

1. Mowing of perennial grasses (alfalfa) for haylage by the self-propelled mower MacDon (operating width – 6 m). Making of the rolls for drying the cut mass
2. Collection of the dried rolls by the rack-roll makers Dominator Tonutti (operating width – 6 m) in one roll

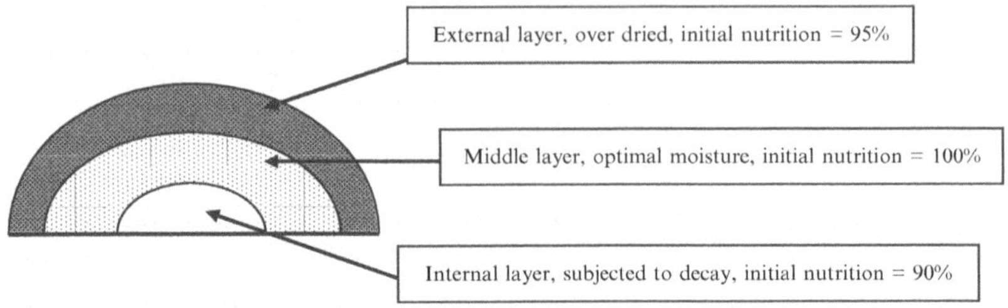

Fig. 11.2 Structure of haylage roll, longitudinal section

3. Picking up and grinding of cut mass by the UEM-280 with loading of gridded mass in the truck aggregated to the tractor MTZ-82
4. Transportation of the haylage in the truck to the specially equipped haylage pit
5. Unloading of haylage into haylage pit, compaction by the tractor MTZ-82
6. Close of haylage pit by the water and airtight film

The simplicity of the haylage harvesting system represents its main asset. Among the disadvantages, the following could be highlighted:

1. When rolls are being made for drying, the whole mass is drying unevenly, and as a result, there is the loss of nutrition of haylage. The scheme of roll's nutrition is presented in Fig. 11.2:

   As it can observe from Fig. 11.2, the total nutrition of haylage in the result is 85% in average from the beginning.
2. During transportation of haylage in a truck, the haylage mass can be penetrated by various harmful substances, microorganisms and inimical bacteria, which can negatively affect the health of animals and their productivity.
3. During compressing of haylage by the tractor, gasoline and oil, dirt and other various harmful substances can also permeate haylage, which also affects the future quality of the feeds and animals' health.
4. During the covering of haylage with protective film, the film itself can also become damaged and can suffer the infiltration of pathogenic microbes and harmful substances, as well as water penetration during winter, which negatively affects the quality of haylage. In addition, covering the haylage pit with the film does not guarantee the sealed conditions for storage, what leads to the damage of feeds.
5. When haylage is used during the winter and spring, it goes to an unsealed haylage pit, and consequently air and water permeate the haylage, leading to the feed decay.
6. The entire harvesting process takes around 5 days.

11.4.2 *Rationale of Managerial Decision and Description of Proposed Harvesting System*

Schematically, the reasons and results of reduction of haylage quantity and quality are presented in Fig. 11.3.

As it can see from Fig. 11.3, low quality of harvesting haylage leads to the complexity of problem, which, finally, leads to the economic instability of a company. The loss of quality and quantity is presented in more detail in Fig. 11.4.

As the result of aforementioned negative factors, the haylage loses its nutrition, which is what can lead to the decline of the amount of revenues from haylage realization because of the low quality and to the decline of productivity of cows because of bad quality feed (LBR Agromarket 2017; Labocky 2013; Krestiansky dom 2015).

For minimizing such negative impact of these factors on the quality of harvested haylage, and as the result on the health of animals and their productivity, we propose to review the harvesting system and implement new agricultural machinery according to the requirement of modern harvesting process. This process

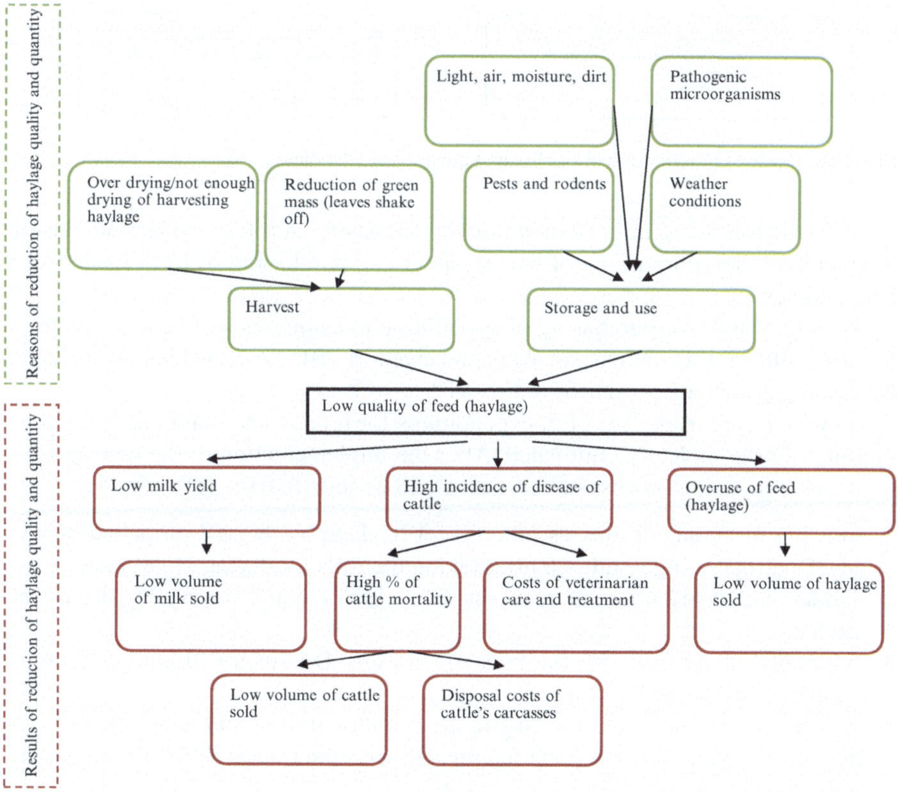

Fig. 11.3 Scheme of reasons and results of quality and quantity decline of harvesting haylage

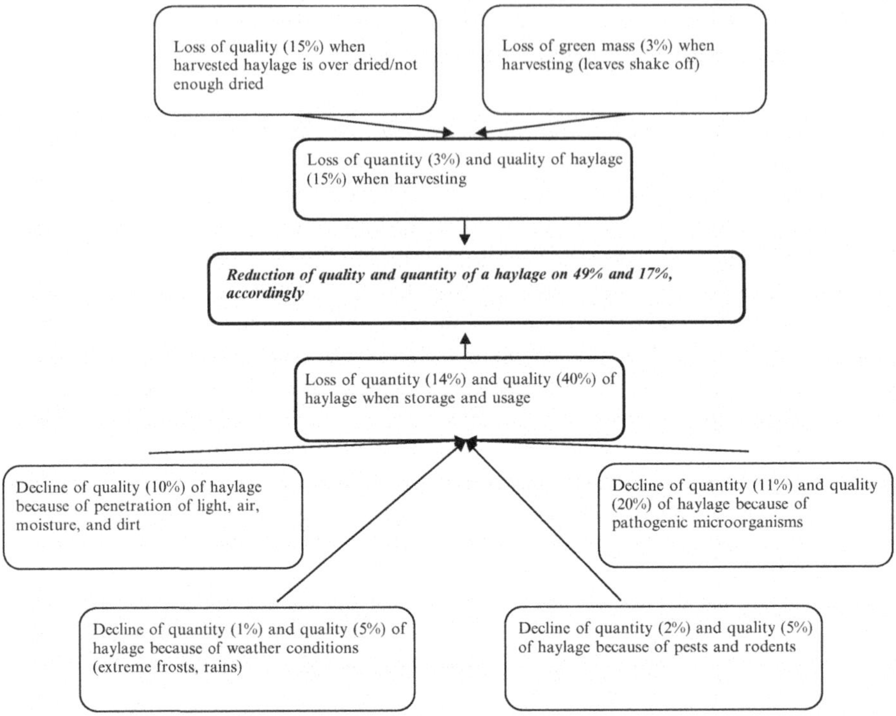

Fig. 11.4 Reasons of reduction of quality and quantity of a haylage in %

includes the following initial expenditures: purchasing of rotary tedder, purchasing of chamber baler, purchasing of bale wrapper and purchasing of front loader with bale grabber.

Required funds for purchasing of agricultural machineries will be invested from the net profit of JSC "Novoazovskoe", received in 2015. Advantages of proposed harvesting system are presented in Fig. 11.5.

As we can see from Fig. 11.5, expenditures for repairs and maintenance will be minimized or completely eliminated. After the implementation of the new agricultural technology, the process of harvesting will be look like the following:

1. Mowing of perennial grasses (alfalfa) for haylage by the self-propelled mower MacDon (operating width – 6 m). Making the rolls for drying of cut mass
2. Tedder and distribution of rolls on the field for equal drying by the rotary tedder
3. Assembly of the rolls by the rack-roll makers Dominator Tonutti (operating width – 6 m) in one big roll
4. Picking up, compression, forming in the cylindrical form and wrapping of bales by the grid with chamber baler aggregated with the tractor MTZ-82, unloading of wrapped bales on the field

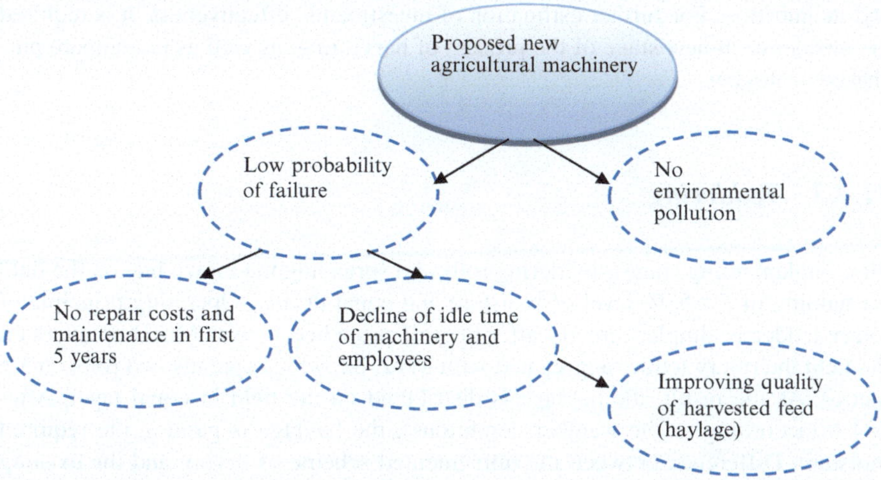

Fig. 11.5 Effect after implementation of new agriculture machineries for haylage harvesting process

5. Picking up and wrapping of bales by the moisture and airtight film agrostretch by the bale wrapper aggregated with the tractor MTZ-82, unloading of packed bales on the field
6. Picking up of packed bales by the front loader with bales grabber aggregated with the tractor MTZ-82 and uploading of them into the tractor's truck aggregated with the tractor MTZ-82
7. Delivery of packed bales to the store and unloading with the front loader with bales grabber aggregated with the tractor MTZ-82
8. Storage of packed bales of haylage, to be used when the need arises

Advantages of this type of harvesting are the following:

Reduction of the of harvesting period from 5 to 2 days

Absence of necessity to make the special waterproof haylage pit

Total sealing of production, protection from penetration of air, moisture, harmful substances and microorganisms and direct sunlight

Keeping of haylage with moisture of 50–55% in which growth of microorganisms in haylage is impossible

Convenience during transportation and storage of production

In addition, the reason of selection of these harvesting and storage processes of feed is an increased keeping of nutrition in relation to tranche technology (Malinin 2013; Ukrprolife 2012; Zootehnikoff.ru 2014).

The parameters of packed haylage are exceeding the analogical parameters of haylage in tranches, what indicates a much higher level of preservation of the feed and its nutrition. For further estimation of investments' effectiveness, it is required to consider each new stage of the process of harvesting, as well as to compare purchased analogues.

11.4.3 Teddering

First implementing stage – teddering rolls and spreading the mown hay on the field for gaining of 50–55% level of moisture and equal drying. Operating principle of rotary tedder is simple: directly after the self-propelled mower MacDon enters on the field the rotary tedder aggregated with MTZ-82, which spreads two rolls on 5.5 metres. As the result, the haylage is distributed on the field in equal layer. After 2–4 h (according to the weather conditions), the haylage is gaining the required moisture. Difference between the implemented scheme of drying and the existing one is presented in Fig. 11.6.

As the result of spreading of feed on the soil, the daylight goes through the layer of haylage and equally dries it. After the gaining of required moisture (determined by the moisture tester) of the haylage, the haylage is picked up into a big roll for further packing. Further, it is necessary to select one of the rotary tedders, presented in the market of Omsk region. Based on the data collected, it was decided to purchase the rotary tedder SIP-Spider 555. This type of rotary tedder was chosen based on main indicators, such as productivity and price, because the other parameters are almost similar.

11.4.4 Bales Forming: Wrapping

Second implemented stage – picking up of a bale, compression, forming into cylindrical form, wrapping with the grid. The operating scheme of chamber baler is simple: chamber baler aggregated with MTZ-82 with the special pins picks up the rolled haylage and pits it into the barrel, where it is compressed and gets the cylindrical form. Further, the bale is wrapped with the plastic grid and unloaded to the field.

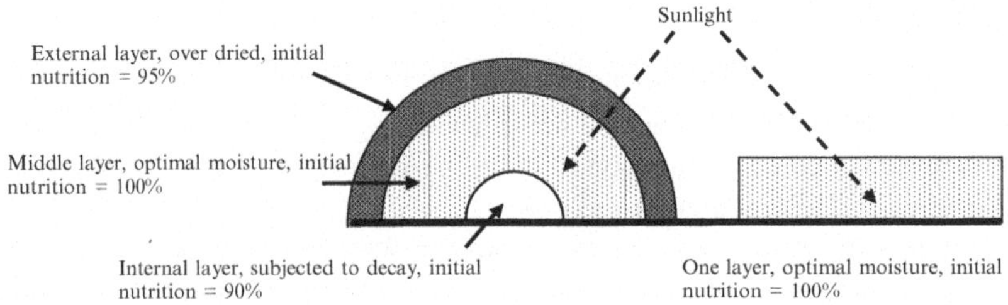

Fig. 11.6 Difference between implemented scheme of drying and existing one

Further, it is necessary to select one of the chamber balers, presented in the market of Omsk region. Based on the data collected, the decision was to purchase the chamber baler Metal-Fach Z562. This type of chamber baler was chosen based on main indicators, such as productivity and price, because the other parameters are almost similar.

11.4.5 Packing of Bales: Uploading/Unloading

Third and fourth implemented stages – picking up of wrapped bale, packing it into the agrostretch, unloading on the field, picking up bales and uploading to the tractor's truck and unloading into storehouse by the bale grabber. Additionally, it is necessary to select one of the bale wrapper presented in the market of Omsk region. Bearing in mind the data, it was decided to purchase the bale wrapper Sipma OS 7510. This type of bale wrapper was chosen based on main indicators, such as productivity and price, because the other parameters are almost similar, namely, appearance of packed bale and the process of packing.

When choosing the front loader, PSN – 1 with bale grabber ZR-1 "Bobruiskagromach" was purchased, because this kind of machinery is the only one sold in the market of Omsk region. The cost of bale grabber with front loader is 399,900 and 203,360 thousand rubles, accordingly.

To sum up, it can be conclude that after the implementation of all the proposed types of agricultural machinery, the company will have a complete process of harvesting (according to modern standards). Advantage of renewable feeds obtained against old ones is its quality, namely, optimal moisture of haylage, absence of any harmful substances in packed product and preserving of nutrition during a long time. Total amount of purchasing machinery is the following:

$$409\,200 + 886\,600 + 328\,600 + 399\,900 + 203\,360 = 2\,227\,660 \text{ rubles.}$$

Economic efficiency of implementation of this type will be presented in the next subsection.

11.4.6 Economic Evaluation of Implemented Measures for Increasing the Economic Sustainability in JSC "Novoazovskoe" of Omsk Region

Considering the level of innovative development of JSC "Novoazovskoe", specific of production activity, financial and economic parameters of enterprise and sizes and structure of production, it is necessary to make an economic estimation of the proposed measures.

Table 11.5 Volume of lost haylage during harvesting, storage and use for 2013–2015

Parameter	Year	2013	2014	2015
Loss of quantity				
Initially harvested mass, centners		127,021	113,121	121,034
Loss of quantity	In average in %	In 100 kg	In 100 kg	In 100 kg
On harvesting	3%	3811	3394	3631
Harvested, centners		123,210	109,727	117,403
Loss due to extreme weather condition	1%	1232	1097	1174
Loss due to pests and rodents	2%	2464	2195	2348
Loss due to pathogenic microorganisms	11%	13,553	12,070	12,914
Total losses, centners		21,060	18,755	20,067
Total losses, %		17%	17%	17%
Remain, centners		105,961	94,365	100,967
Loss of quality				
Harvested, centners		123,210	109,727	117,403
Loss of quality	In average in %	In 100 kg	In 100 kg	In 100 kg
On harvesting	15%	18,482	16,459	17,610
Harvested good haylage, centners		104,729	93,268	99,793
Loss due to air, moisture, light	10%	10,473	9327	9979
Loss due to extreme weather condition	5%	5236	4663	4990
Loss due to pests and rodents	5%	5236	4663	4990
Loss due to pathogenic microorganisms	20%	20,946	18,654	19,959
Total losses of good haylage, centners		60,373	53,766	57,527
Total losses of good haylage, %		49%	49%	49%
Remains of good haylage, centners		62,837	55,961	59,876
Volume of good haylage in total mass, %				
Good haylage in remains		59%	59%	59%

It is required to start the calculation of economic efficiency by providing the data of amount of the produced haylage and its consumption during the year.

Based on previous information, we can calculate remains of good haylage in the total mass of harvested haylage and specify the losses due to the impact of each negative factor. Data for the analysis of lost haylage were obtained from the chief agronomist according to his survey for other 10 years.

Based on Table 11.5, we can say that during the harvesting, the haylage loses its quality in 15% and quantity in 3%. Additionally, when stored and used, the haylage loses its quality in 40% and quantity in 14%.

After the implementation of the new harvesting system, the loss of quantity and quality on the stage of harvest will be 1% and 5% of total mass of haylage, accordingly. When stored and used, the haylage will lose 6% of quantity and 13% of quality. Based on that statement, we can calculate the volume of good haylage in total mass of harvested haylage. Calculations will be performed based on the data covering the years from 2013 to 2015 to giving the example of what could be if the existing harvesting system would be replaced by the modern one. Results are in Table 11.6.

Table 11.6 Proposed results after implementation of new harvesting system in 2013–2015

	Year	2013	2014	2015
Loss of quantity				
Initially harvested mass, centners		127,021	113,121	121,034
Loss of quantity	In average in %	In 100 kg	In 100 kg	In 100 kg
On harvesting	1%	1270	1131	1210
Harvested, centners		125,750	111,989	119,824
Loss due to extreme weather condition	0%	0	0	0
Loss due to pests and rodents	1%	1258	1120	1198
Loss due to pathogenic microorganisms	5%	6288	5599	5991
Total losses, centners		8815	7851	8400
Total losses, %		7%	7%	7%
Remain, centners		116,935	104,139	111,424
Loss of quality				
Harvested, centners		127,021	113,121	121,034
Loss of quality	In average in %	In 100 kg	In 100 kg	In 100 kg
On harvesting	5%	6288	5599	5991
Harvested good haylage, centners		119,463	106,390	113,832
Loss due to air, moisture, light	3%	3584	3192	3415
Loss due to extreme weather condition	1%	1195	1064	1138
Loss due to pests and rodents	2%	2389	2128	2277
Loss due to pathogenic microorganisms	7%	8362	7447	7968
Total losses of good haylage, centners		21,818	19,430	20,789
Total losses of good haylage, %		17%	17%	17%
Remains of good haylage, centners		103,933	92,559	99,034
Volume of good haylage in total mass, %				
Good haylage in remains		88%	88%	88%

As we can see from Table 11.6, there is 88% of good haylage remaining in total mass of the haylage. Therefore, based on that information, we can calculate the denied gross profit, received from selling of surplus of haylage received from the difference of required volume of haylage: from item "remain of good haylage" in Table 11.6, we subtract the item "remains of good haylage" in Table 11.5. Results are in Table 11.7.

As we can see from Table 11.7, after the implementation of the new harvesting system, accumulated proposed gross profit by the end of 2015 could be more than 42 million rubles (Flagma, 2017). That means that JSC "Novoazovskoe" denied more than 42 million rubles of additional potential gross profit, because it did not implement a modern harvesting system. This result demonstrates that the new harvesting system is reasonable for implementation.

Additional, it is necessary to summarize all data to notice the final results of the new harvesting system implementation. The year 2015 will be taken as the based year. Initial expenditures for purchasing of new machineries will be made in the beginning of next year (January to February), so the results of new harvesting

Table 11.7 Financial results from selling the denied surplus

Parameter	2013	2014	2015
Remain of good haylage in new system, centners	103,933	92,559	99,034
Remain of good haylage in old system, centners	62,837	55,961	59,876
Haylage for sale, centners	41,096	36,598	39,159
Price per centners, rubles	450	500	550
Proposed revenues, rubles	18,493,027	18,299,240	21,537,308
Accumulated proposed revenues, rubles	18,493,027	36,792,267	58,329,575
Cost of haylage, rubles/centner	104,72	183,90	132,34
Total cost of haylage, rubles	4,303,533	6,730,460	5,182,268
Accumulated cost, rubles	4,303,533	11,033,993	16,216,261
Proposed gross profit, rubles	**14,189,494**	**11,568,779**	**16,355,040**
Accumulated proposed gross profit, rubles	14,189,494	25,758,274	42,113,314

Table 11.8. Initial expenditures and financial results of new harvesting system implementation

Parameter	2015	2016	2017	2018	2019	2020
Haylage for sale						
Surplus, centres	0	39,159	39,159	39,159	39,159	39,159
Haylage for sale of remains, %	0	20%	30%	40%	50%	60%
Haylage for sale of remains, centners	0	7832	11,748	15,663	19,579	23,495
Costs and revenues						
Initial expenditures on machinery, rubles	0	2,227,660	0	0	0	0
Cost of agrostretch, rubles	0	100,000	100,000	100,000	100,000	100,000
Cost of use and maintenance, rubles	0	0	12,000	12,000	12,000	12,000
Cost of haylage, rubles/ centner	132	132	132	132	132	132
Total cost of haylage, rubles	0	1,036,454	1,554,680	2,072,907	2,591,134	3,109,361
Accumulated cost, rubles	0	1,036,454	2,591,134	4,664,041	7,255,175	10,364,536
Price of haylage, rubles/ centner	550	550	550	550	550	550
Revenues, rubles	0	4,307,462	6,461,192	8,614,923	10,768,654	12,922,385
Accumulated revenues	0	4,307,462	10,768,654	19,383,577	30,152,231	43,074,616
Total gross profit, rubles	**0**	**943,348**	**4,794,512**	**6,430,016**	**8,065,520**	**9,701,024**
Accumulated total gross profit, rubles	0	943,348	5,737,860	12,167,876	20,233,396	29,934,420

system implementation were already obtained in the end of harvesting season and year 2016. Calculation will be given for 5 years (from 2016 to 2020). Results are in Table 11.8.

As we can see from Table 11.8, JSC "Novoazovskoe" will generate profit from selling the remains of haylage. Total accumulated gross profit for 5 years is more than 29 mln rubles. This fact means that it is much more profitable and reasonable for the company to implement the proposed harvesting system, than to use the existing one.

Summing up, it can be asserted that development of an enterprise in the branch of haylage selling is one of the prospective directions. One reason for that to happen has to do with the lack of enterprises in Omsk region, specializing on haylage production. Result of this lack is that there are no strong competitors in the market, like in the grain or milk markets. Because JSC "Novoazovskoe" is one of the leaders in milk production, the company should implement the production of new feeds or improve the existing one. As the result of implementation of new feeds, the company can take its niche onto new markets of agricultural production; entering this market with the production meets the modern requirements of quality of haylage and conditions of storage. Further, for JSC "Novoazovskoe" it is possible to enter in the market of Siberia, Russia and foreign countries.

11.5 Conclusions, Limitations and Future Research Lines

Based on implemented analysis of the definition of "economic sustainability of enterprise", it is possible to conclude that this has a complicated meaning, which includes many components, such as production and technical, commercial, organizational, innovation, financial, social, industrial and regional sustainability. One of the main features of economic sustainability increase in agricultural enterprise is diversification of crops (including livestock), harvesting and cultural practices to enhance the biological and economic stability of the farm.

By making the analysis of activity of JSC "Novoazovskoe", the following information was checked:

- Company's balance sheet increased from 319,543 thousand rubles to 368,460 thousand rubles from 2013 to 2015, accordingly.
- Company's net profit decreased from 32,167 thousand rubles to 19,787 thousand rubles from 2013 to 2015, accordingly.
- Company's total square of agricultural lands increased in 82 hectares from 11,118 hectares in 2013 to 11,200 hectares in 2015, where 100% of it is arable.
- Company's total number of means of production decreased from 339 units in 2013 to 207 units in 2015 but in monetary term increased from 91,177 thousand rubles in 2013 to 100,086 thousand rubles in 2015.
- Number of cattle decreased in 8 units from 2668 units in 2013 to 2660 units in 2015 and in monetary term increased from 88,060 thousand rubles in 2013 to 104,335 thousand rubles in 2015. The number of milking cows was the same for 2013–2015 and equal to 1023 and in monetary term increased in 10,281 thousand rubles from 47,821 in 2013 to 58,102 thousand rubles in 2015.

- Number of employees decreased from 294 to 235 units from 2013 to 2015 accordingly and in monetary term salary increased from 68,924 thousand rubles in 2013 to 73,336 thousand rubles in 2015.
- Total volume of milk produced increased from 48,707 centners in 2013 to 54,272 centners in 2015 and in monetary terms increased from 87,670 thousand rubles in 2013 to 199,495 thousand rubles in 2015.

Based on the implemented SWOT analysis, it was examined that the weaknesses include low quality of harvested feeds because of big loss of harvested feeds as the result of improper and old-style process of harvesting and storage, unacceptable conditions of storage of haylage and its loss of more than half of nutrition value, lack of modern equipment for feed harvesting, storage and delivery.

The results showed that after implementation of a new haylage harvesting system, the company increased the volume of a good haylage in total harvested mass of haylage up to 80% (achieved 83%); and implementation of the new haylage harvesting system can bring additional revenues in total income for a company.

To improving that problem, it is proposed the implementation of a new haylage harvesting system, which includes changing the harvesting process and purchasing new required agricultural machineries. These machineries include rotary tedder, chamber baler, bale wrapper and front loader. The total cost of purchasing agricultural machineries is 2,227,660 rubles. After implementation of modern haylage harvesting system, the quality of harvested haylage increased up to 83% of good haylage in total mass and 88% in remains.

After realization of a surplus of haylage, JSC "Novoazovskoe" will receive additionally 29,934,420 rubles of gross profit, which means that the proposed system is reasonable for implementation.

References

Agrovesti.net. (2017). Technology of harvesting and storage of a haylage. [On-line]. Available: https://agrovesti.net/kormoproizvodstvo/technologiya_zagotovki_i_chraneniya_senazha.html. Accessed 10.06.2017.

Anisimov, Y. (2006). *Management of sustainable development of a company based on innovations and entrepreneurship*. Voronezh: VSTA.

Bulatov, A. (2007). *Economics: Study book*. Moscow: Yurist.

Feenstra, G., Ingels, C., & Campbell, D. (2016). What is sustainable agriculture? [On-line]. Available: http://asi.ucdavis.edu/programs/sarep/about/what-is-sustainable-agriculture/#plant-production-practices. Accessed 10.06.2017.

Flagma.ru. (2017). Haylage for feed. [On-line]. Available: http://kommunarlenin.flagma.ru/senazh-korm-provyalennyh-trav-o3038356.html. Accessed 10.06.2017.

Folomjev, A. (1995). *Sustainability of enterprise on the market*. Moscow.

Goldstein, G. (2003). *Basics of management: Study book* (2nd ed.). Taganrog: TRTU.

Golovko, E. (2013). Definitions of sustainable system. *Journal "Young scientist"*, *5*, 283–283.

Koryakov, A. (2012). Economic sustainability of enterprise development: Classification and key factors. *Journal "Modern Science: actual problems of theory and practise"*, *3*, 5–10.

Krestniaksky dom. (2015). Haylage in package. [On-line]. Available: http://www.kd59.ru/tehnologiya-senazh-v-upakovke. Accessed 10.06.2017.

Kuzmenko, O., & Gritcenko, E. (2016). Economic entity of sustainability of agriculture production. *Journal "Scientific and methodological electronic journal "Koncept""*, *40*, 35–38.

Labocky, I. (2013). Feeds from a hose. *Journal "Belarus agriculture"*, 124–129.

Laibert, T. (2011). *Economic sustainability: Entity, content and factors*. Ufa: USSTU.

Lakshina, O., & Chekmareva, E. (2009). Analysis of financial stability: Practice and methodology. *Journal "Money and credit"*, *3*, 24–29.

LBR Agromarket. (2017). Haylage in package – Make the most nutritious feed. [On-line]. Available: https://www.lbr.ru/company/events/tehnologija-zagotovki-i-hranenija-senazha-v-upakovke/. Accessed 10.06.2017.

Malinin, I. (2013). What feeds can tell to us. *Journal "Agricultural News"*, 70–73.

Marks, K., & Engels, F. (1984). *Capital: Critique of political economy*. Moscow: Politizdat.

OmskPortal.ru (2017). Azovsky German National District [On-line]. Available: http://azov.omskportal.ru/ru/municipal/localAuthList/3-52-201-1/officialsite.html. Accessed 10.06.2017.

Osita, I., Onyebuchi, I., & Nzekwe, J. (2017). Organization's stability and productivity: The role of SWOT analysis an acronym for strength, weakness, opportunities and threat. *International Journal of Innovative and Applied Research*, *2*(9), 23–32.

Ukrprolife.com. (2012). Feeds: Haylage, silage. [On-line]. Available: http://ukrprolife.com.ua/krupnyj-rogatyj-skot-2/korma-silos-senazh/. Accessed 10.06.2017.

Zootehnikoff.ru. (2014). Haylage. [On-line]. Available: http://www.zootehnikoff.ru/senazh/. Accessed 10.06.2017.

Chapter 12
Innovation Management in Portuguese and Russian Agricultural Companies

Mikhail Kopeykin, Sofia Cardim, Vitaly Aleshchenko, and Frederico Branco

12.1 Introduction

Nowadays, developing innovation programs in enterprises is one of the ways to increase the volume of production, to improve the quality of goods and services, and last but not least, to survive on the market. Innovative activity became an important part of the management of strategic decisions. For agricultural companies, the implementation and use of innovations is also an important part of their activity, because using and following new technologies can lead to a greater success for the business itself.

The most popular innovations in agriculture right now are related to the optimization of the cost structure of activity and the improvement of the quality of products. Under optimization of cost structure, the decline of the cost of resources or the removal of non-required costs is taken into consideration. One of the actual trends is producing electricity through company's own sources, namely, alternative energy such as wind, sun, biofuel, etc. Among all the renewable sources, the sun is most preferable, because solar panels can be installed on existing constructions, without providing additional separate squares. Implementing of such a technology leads to

M. Kopeykin (✉) · V. Aleshchenko
Omsk State Agrarian University, Institutskaya ploshad, Omsk, Russia
e-mail: vv.aleschenko@omgau.org

S. Cardim
Department of Economics and Management, School of Technology and Management, Polytechnic Institute of Braganca, Campus de Santa Apolónia, Braganca, Portugal
e-mail: sofiacardim@ipb.pt

F. Branco
INESC TEC and UTAD, School of Science and Technology, University of Trás-os-Montes e Alto Douro, Vila Real, Portugal
e-mail: fbranco@utad.pt

M. Peris-Ortiz et al. (eds.), *Knowledge, Innovation and Sustainable Development in Organizations*, Innovation, Technology, and Knowledge Management,
https://doi.org/10.1007/978-3-319-74881-8_12

a decrease in the cost of electricity, generating additional funds and sustaining the opportunity to invest in other activities or branches.

The main objective of this research work is finding theoretical statements and validation measures of an innovative activity increase in Russian and Portuguese enterprises, and more concretely, considering the main problem and ways of increasing innovative activity of enterprises in general, particularly agricultural enterprises, analyzing the activity of Russian and Portuguese enterprises, analyzing ways of increasing the innovation activity of enterprises, and justifying measures of innovative activity increase in Russian and Portuguese enterprises by implementing a renewable source of energy, namely, solar panels. The results of the implemented projects will be compared.

Concerning the objects of observation, we will review the agricultural production cooperative "Bolshevik" of Moskalensky District of Omsk region of the Russian Federation and "Company B" LDA of Bragança Municipality of the Portuguese Republic. The object of research is the innovative activity of APC "Bolshevik" and of "Company B" LDA. The subject of research is the implementation of renewable sources of energy, as a way of increasing the innovative activity of enterprises. This research uses as input data the bookkeeping and managerial statement of APC "Bolshevik" and "Company B" LDA for 2013–2015; open Internet sources related with financial, statistical, managerial, and other information; libraries of OmSAU, IPB, and international databases and libraries; scientific and specialized online and regular journals of both countries, as well as international ones; data from governmental portals and sources related with the theme of investigation; and information from the companies-producers of renewable sources of energy. All the data is presented in euros, and the data for the Russian enterprise is converted from rubles to euros. Exchange rate is equal to 1 euro = 74,39 rubles (Investing.com 2017).

12.2 Theoretical Bases of Innovative Activity Management in an Agricultural Complex

12.2.1 Concept and Essence of Innovation: Types, Classification, and Functions of Innovations

During the last few years, most economic and social trends that existed for centuries were broken. The best examples for that are decrease of price for energy sources such as crude oil of all marks (average decrease is of more than 70%), as well as for raw materials, such as aluminum (over 40%), lead (over 40%), and gold (over 30%) (Investing.com 2017); reduction in prices of flights into space such as that of Space X which made the world's first successful return of the carrier-rocket's first stage (Spacex 2017); classic manned vehicles being replaced by unmanned vehicles (Google 2017); practical experience on growing new human and animal organs, which can give hope to people who are diagnosed with chronic illness or for any reason have lost one of their organs (FGBE "FSCTAO named after V.I. Shumakov" of Health Ministry of Russia 2017); and widespread use of biofuel instead of traditional energy sources, as well as many, many other examples (Bioenergy International 2017; Choren 2017). Such serious changes are playing a main role in the development of human civilization. Even though they are different through their fields of activity and dimensions, all these trends have something in common – they are all innovations.

Concerning the implementation of new ideas, innovation provides the improvement in the efficiency in several items, specifically not only in manufactured products but also in processes (Bernar et al. 1997). "It's the result of human intellectual activity, his imagination, creative process, discoveries, inventions, and rationalization" (Abalkin 1999). Innovation can be found when a new product arrives to the market or when an existing product is improved or produced more efficiently. The very notion of innovation was first introduced in the scientific researches of the nineteenth century, but a new life was given to this concept in the beginning of twentieth century in the scientific work of the Austrian economist J. Schumpeter in the analysis of "innovative combinations" and changes in the development of economic systems (Academic 2017).

In accordance with Schumpeter, "the fundamental impulse that sets and keeps the capitalist engine in motion comes from the new consumer goods, the new methods of production or transportation, the new markets, the new forms of industrial organization that the capitalist enterprise creates" (Schumpeter 1976).

The economy is agitated by innovation, which is the engine of economic development. In fact, without entrepreneurship and innovation, progress could not be achieved (Schumpeter 1934).

Furthermore, in accordance with Schumpeter (1934, p. 66), the concept of "new combination" covers the following five cases: "(1) The introduction of a new good – that is one with which consumers are not yet familiar – or a new quality of a good. (2) The introduction of a new method of production, that is one not yet tested by experience in the branch of manufacture concerned, which needs by no means be founded upon a scientifically new discovery and can also exist in a new way of handling a commodity commercially. (3) The opening of a new market that is a market into which the particular branch of manufacture of the country in question has not previously entered, whether or not this market has existed before. (4) The conquest of a new source of supply or raw materials or half-manufactured goods, again irrespective of whether this source already exists or whether it has first to be created. (5) The carrying out of the new organization of any industry, like the creation of a monopoly position (for example through trustification) or the breaking up of a monopoly position."

Innovation is the result of investing in intelligent solutions for development and obtaining a new knowledge, not basing yourself on previously used ideas for updating several and distinctive domains of life and following the subsequent process of implementation (production) of the new knowledge, with a fixed additional value (Economic and law 1998; Zaichenko 2011).

Thus, the required process is investments–development–implementation process–obtaining of qualitative improvement. Innovation is a process or result of the process, which (Azgaldov and Kostin 2008) uses partially or completely the protectable results of intellectual activity; ensures the production of patentable products; ensures the production of goods and/or services, with its quality corresponding to world standard or higher; and achieves high economic efficiency in the production or consumption of the product.

Because of the diversity of changes in a person's life, it is not equitable to compare the degree of their influence; they must be classified according to some common characteristics, namely (Kalenskaya 2012), technological innovations, ecological innovations, economic innovations, sociopolitical innovations, and spiritual innovations.

Classification of innovations can be implemented by different schemes and by using various classification features. In economic literature we can find various approaches to the classification of innovations, as well as to identifying criteria.

Innovations are classified by the following features (Kalenskaya 2012):

1. Importance: basis, improving, and pseudo-innovations.
2. Directivity: replacing, streamlining, and extending.
3. Place of realization: industry of origin, industry of implementation, and industry of consumption.
4. Depth of changes: regeneration of initial methods, modifying of quantity, regrouping, adaptive variations, new form, new generation, new species, and new genus.
5. Developer: developed by the enterprise or by external forces.
6. Distribution scale: for creation of a new industry and for use in all industries.
7. Position in production process: major product and technological or complementary product and technological.
8. Features of satisfying needs: new needs or existing needs.
9. Degree of novelty: based on scientific discoveries or on new methods applied to previously discovered phenomenon.
10. Period of market entry: innovation leaders or innovation followers.
11. Reason of appearance: reactive and strategic.
12. Field of application: technical, technological, organizational and managerial, informational, social, etc.

By the position of innovation in a system (in a company), the following can be selected (Kalenskaya 2012): innovations "on the input" (changes in the selection of raw materials, fabrics, machineries and equipment, information, etc.), innovations "on the output" (products, services, technologies, information, etc.), and innovations of enterprise's system structure (managerial, production, technological).

Regardless of the method of classification of innovations, it's necessary to consider the fact that not every novelty is an innovation. In order to be called an "innovation," the implemented novelty should possess a number of properties (Kalenskaya 2012): scientific-technical novelty (all introduced products, services, and processes should be new (or sufficiently upgraded)), production application (there should be presence of certain conditions and opportunities to implement this novelty into a specific product), and commercial feasibility (the created product should meet the market demand, be commercially feasible, and, after all, generate profit to manufacturer).

Innovation is the performed result on the market, accepted from the investments of the capital into a new product of operation (technology, process). Based on this fact, we can say that innovation is implementing the following functions (Kalenskaya 2012): reproductive function, investment function, and stimulating function.

Reproductive function (Kalenskaya 2012) means that innovation is an important source of financing to extend production. Cash revenue obtained from selling the innovation on the market makes entrepreneurial profit, which turns into a financial resource and simultaneously into the measure of efficiency of the innovation process. It can be transferred to increasing the volume of production, in trading, investment, innovation, and other financial activities, which sums up the content of the reproductive function of innovations.

Profit obtained from selling an innovation can be used in different ways, as well as the capital. This capital can be directed to the financing of all investments, or particularly the new kinds of innovations, which is the content of the *investment function* (Kalenskaya 2012) of innovation.

Generating profit from selling innovative products is the main goal of each commercial enterprise. This is the incentive for any entrepreneur to make new innovations; it motivates him to dig deep in the issue, improve the organization of all marketing activity, and use much more modern approaches to financial management (reengineering, brand-strategy, benchmarking, etc.); all of this is the content of the *stimulating function* (Kalenskaya 2012) of innovation.

Summarizing all those mentioned above, we can notice that in scientific literature, the meaning of "innovation" started being used not such a long time ago, and scientists still do not have the common definition for this term. It is related to the field of application of innovations – it covers all human life – and couldn't be considered separately from a specific field of activity. Nevertheless, besides variety, an essential feature of any innovation is scientific and technical novelty, as well as economic feasibility.

12.2.2 Innovative Potential of a Company: Innovative Management in Agricultural Companies and Features of Innovative Activity in Agribusiness Industry

Innovative potential (Lukov et al. 2012) is described as the organization's ability to achieve the objectives through the implementation of innovative projects. For the convenience of analyzing innovative potential, the projects are presented in descending order of effectiveness; effects and costs are represented as the cumulative sums. The amount of innovative potential is determined by the available research, design, technological organizations, pilot plants, experimental polygons, educational institutions, personnel, and technical resources of these organizations (Lukov et al. 2012).

Innovative management in agriculture companies can help to solve the difficulties in the organization of the complete and uninterruptible process of agriculture production and delivery of a product by using (for instance) the supply chain management, as the example of solution (Hussain et al. 2015). The relation between technological innovations and agriculture results has several benefits, which can include, among others, time and cost advantages. Moreover, since agricultural products are perishable and have a short period of consumption possibility (in most cases), they need to be conserved, with recourse to technology (cold and freeze storage), or even processed using several operations. Technology can be observed, not only in the process itself but also in the cost reduction of it (Tseng et al. 2011; Hussain et al. 2015).

The advantage creation is observable not only in all the processes associated to preservation and transportation but also in the concept of "time for space" which in many cases constitutes a great competitive advantage (Gunasekaran et al. 2006; Hussain et al. 2015).

The market for agricultural products is not characterized by rapid transfer of information between all its agents. Competitors, collaborators, market demand, and seasonality are not always well aligned, and sometimes the market adjustment does not take place swiftly (Garbi 2002). Innovation and technology would help not only the agents but also to reduce spoilage and waste and therefore the maintenance of natural resources and sustainability. Likewise, the information exchange would help to avoid uncertainty or risk leading to cost reduction (Ryan et al. 2000; Hussain et al. 2015).

We are looking to define the value of development if agribusiness industry and the agricultural sector have new, innovative investment policies, on the basis of providing highly developed agriculture and food security – modern production processes, technologies, and equipment.

Along with a quite problematic economical statement of the Russian agribusiness industry and its enterprises, a complicated economical and geopolitical situation formed in the last 3 years; in recent years, the innovation process has gradually become more active. It especially applies to leading agriculture organizations, as well as to food industry companies, which are intensively acquiring new scientific achievements and innovations for their production systems. And in the conditions of the imposed food embargo, innovation activities began to develop in small and medium agricultural enterprises, which had a more conservative approach of making business.

However, despite many difficulties, science is developing in a positive direction, improving communication for both scientists and agro-entrepreneurs from different countries. Among the top projects of the world's agricultural industry, which are in their final phase of development or are already performing, we can review the following:

- Agricultural robot (or agrobot). The main field these robots can be applied to is plant growing – the period of harvest for grain and leguminous plants – but they can also pick fruits and autonomously drive tractors/sprayers. Besides, robots can also implement pruning, weeding, plowing, watering, and monitoring of all agricultural activities of a company (Case 2017; The Robot Report 2017). In animal breeding, robots are used for feeding, watering, milking, cleaning, castration, shearing (sheep), collecting eggs (on egg plants), and waste disposal. In addition, agrobots are widely used in the field of agricultural products' processing and packing.
- Precision farming. On the basis of the scientific concept of precision farming, there are some ideas about the existence of non-homogeneity within a single field. For estimating and detecting these non-homogeneities, the newest technologies are being used, such as Global Positioning Systems (GPS, GLONASS), special sensors, air photos and photos from satellite, as well as special programs for agricultural management, based on the geo-information systems (GIS). Collected data is used for planning of the seeding process, calculation of fertilizers norms and plant protection agents (PPA), more accurate forecastings of crop yield, and financial planning (NASA Earth Observatory 2001).
- Vertical farms. It's a general name for a highly automated agriculture industry, placed in specially designed high-rise buildings, therefore the name. The main difference between vertical farms, traditional greenhouses, and livestock farms is an intensive approach to the use of the territory by using a vertical multitiered placement of plants. In fact, the farm is a multistory greenhouse. The prerequisite for developing such projects was the constant growth of the world's population, which in the foreseeable future will lead to a shortage of territories for agricultural purposes (Plantagon 2017).

All of the projects mentioned above are basically related to agricultural places where the limit of extensive agriculture was already reached, and there is no possibility to increase the area of agricultural plots (the Netherlands, South Korea); the construction is mainly in height, and transportation costs are large enough for delivery of the fresh produce from agricultural regions to giant metropolitan areas (USA, China).

In the Russian Federation, none of these problems exists, because only 40% of all agriculture plots are used in production, and the cities and metropolitan areas do not have such big dimensions in order to obstruct the relatively cheap delivery of the fresh production (FSSA 2017); as for the Portuguese Republic, none of these problems is a serious obstacle for the further intensive development of agriculture, since a sufficiently large area of non-using agriculture plots allows to do so (INE 2017).

Among the most prospective projects and trends of development for both Russian and Portuguese agribusiness, the implementation of intensive farming is to be noted, through the increase of productivity per unit square through breeding of new species and the introduction of new technologies in crop production. In animal breeding these include breeding of new high-yielding breeds of animals, as well as other unique breeds and implementing of "green energy." This trend can significantly help to reduce the cost of electricity, since enterprises will produce it by using solar or wind energy or by using biofuel. This latter innovation can reduce the costs on acquisition of fuel, because the required amount can be obtained by recycling organic materials, subsequently obtaining biogas (biofuel) and recycled organic waste (fertilizer).

Agriculture and food production are a powerful incentive for the development of many areas of science and in general for most innovation processes, because they completely cover an interrelated system of activity: nature–human–economy. This system uses innovation achievements of all other sectors of science and at the same time provides the resources for their functioning and development (Stukach 2007).

In summary, we can conclude that right now, in Russian and Portuguese agricultural enterprises, innovation and investment activities in agriculture are in their infancy, with only rare interest to leading achievements in top enterprises. However, because of recent events, specifically the extension of the product embargo until the end of 2017, Russian agricultural enterprises obtained the most favorable conditions and opportunities for future development and increase of economic efficiency. For Portuguese enterprises, because of the relatively low cost of the basic production assets and labor, as well as a favorable geographical position at the crossroads of trade routes from Europe, Africa, and North and South America, the opportunity for rapid and qualitative economy growth has emerged, both in general and for agricultural enterprises in particular. The final result of their activities will completely depend on the management ability of agricultural enterprises to receive advantages from new opportunities. However, the first priority for companies is to make an inventory of existing capacities and their optimization and modernization. These processes will have a serious impact on their sustainable development in the future.

12.3 Organizational and Economic Evaluation of the Activity of an Enterprise

12.3.1 Organizational Characteristic of the Russian Enterprise

Further consideration of practical aspects of the organization process of innovative activity in agricultural companies will be implemented based on the data of the Russian company agricultural production cooperative "Bolshevik" (hereinafter – APC "Bolshevik"). This company was selected from a number of agricultural enterprises as the most typical representative of industry; according to the type of organizational structure, APC is common for companies founded in the USSR and has balanced structure of employees, lands, and quality of production; the company is a medium enterprise, which is mostly common for agricultural companies of Omsk region according to the number of employees; the bookkeeping statements and Omsk Statistics Department, as well as data of the company itself, show that it's an innovative company, due to application of modern agricultural technologies and machineries.

APC "Frolova et al (2015)." is characterized by a linear-functional type of organizational structure. The linear-functional management structure ensures such a diversification of managerial work, in which the linear managerial elements provide overall leadership and coordination, and the functional ones consult and develop specific questions (Goldstein 2003). As follows from the aforesaid, the structure of APC "Bolshevik" is linear with direct subordination, balanced in terms of number of services and departments, as well as several linear managers.

12.3.2 Financial and Production Characteristic of the Activities of APC "Bolshevik"

APC "Bolshevik" of Moskalensky District of Omsk region is located in a steppe zone. The climate is typically continental. Unhindered penetration of cold Arctic air masses from the north and dry masses from Kazakhstan and Central Asia leads to sharp and rapid weather changes and to overall instability of the climate. The climate is characterized by dryness, lack of precipitation, and low clouds. The average temperature in January is −19 °C/−20 °C. The average temperature in July is +17 °C/−19 °C. Annual precipitation is 300–430 mm. Stable snow cover forms in late October to early November; the height of it by the end of winter reaches 35–50 cm. Directions of prevailing winds are western and southwestern, and only in summer northern and northwestern winds appear. The positive side of the climate lies in a lot of sunshine and heat during the growing season, which largely compensates the short period of positive temperatures and accelerates the vegetation of plants. The territory is a vast plain with numerous lakes.

For making an estimation of economic and business efficiency of this enterprise's activity, it is necessary to analyze following indicators (Horngren et al. 2012): change of balance sheet structure from year to year, change in financial results from year to year, dynamics of changes in number and amount of current assets, dynamics of changes in area and quality of lands, dynamics of changes in average number of employees, and production of various kinds of agricultural products.

Firstly, we inspected the balance sheets and the income statements of APC "Bolshevik" as the most important and reliable source of information. Vertical and horizontal analyses were conducted in order to get more information about the trends of the enterprise's development. The analysis of balance structure was made for 3 accounting years – 2013–2015. These years were chosen because of unavailability of information in the year 2016. The data was obtained from accounting statements, and a horizontal analysis year-to-year was performed. "Year-to-year" analysis means that all years will be compared in an ascending sequence.

For the further analysis of business activity of an enterprise, it was necessary to analyze the income statement of APC "Bolshevik" for 2013–2015. It was also necessary to make both the vertical analysis and horizontal analysis to get more information about the trends of the enterprise's development, and the analysis of income statement structure was made for 3 accounting years. The data were obtained from accounting statements and presented in Table 12.1 (year-to-year horizontal analysis) (Horngren et al. 2012).

All the items in the income statement were compared with "total revenues," because in this case we can see the structure of revenues and expenditures and their influence on total income. The section "Additionally" shows the provision for a single

Table 12.1 Year-to-year horizontal analysis of APC "Bolshevik" of income statement

Name of item	Year			2014 to 2013		2015 to 2014	
	2013	2014	2015	In amount	% of previous	In amount	% of previous
Revenues							
Plant growing	350,590	304,616	472,948	−45,975	86.9%	168,332	155.3%
Animal breeding	1,660,008	2,225,147	2,519,116	565,139	134.0%	293,969	113.2%
Other goods	36,081	9813	17,933	−26,267	27.2%	8120	182.7%
Other services	11,238	14,317	13,954	3078	127.4%	−363	97.5%
Total revenues	**2,057,917**	**2,553,893**	**3,023,950**	**495,975**	**124.1%**	**470,057**	**118.4%**
Cost of revenues							
Plant growing	185,525	192,475	304,844	6950	103.7%	112,369	158.4%
Animal breeding	1,544,386	1,784,691	2,023,530	240,305	115.6%	238,840	113.4%
Other goods	35,892	9813	17,933	−26,079	27.3%	8120	182.7%
Other services	5646	14,559	9208	8913	257.9%	−5350	63.3%
Total cost of revenues	**1,771,449**	**2,001,538**	**2,355,516**	**230,088**	**113.0%**	**353,978**	**117.7%**
Gross profit							
Plant growing	165,065	112,141	168,103	−52,925	67.9%	55,963	149.9%
Animal breeding	115,622	440,456	495,585	324,834	380.9%	55,129	112.5%
Other goods	188	0	0	−188	0.0%	0	0.0%
Other services	5592	−242	4745	−5834	−4.3%	4987	−1961.1%
Total gross profit	**286,468**	**552,355**	**668,434**	**265,887**	**192.8%**	**116,079**	**121.0%**
Other items							
Other incomes	174,932	152,227	139,940	−22,705	87.0%	−12,287	91.9%
Other expenses	86,236	99,410	93,388	13,174	115.3%	−6022	93.9%
Profit before income taxes	**375,164**	**605,172**	**714,987**	**230,008**	**161.3%**	**109,815**	**118.1%**
Interest income	0	0	0	0	0.0%	0	0.0%
Interest expense	0	0	0	0	0.0%	0	0.0%
Other income (expense), net	1304	0	0	−1304	0.0%	0	0.0%
Net profit	**373,860**	**605,172**	**714,987**	**231,312**	**161.9%**	**109,815**	**118.1%**
Additionally							
Provision for single agricultural tax	22,432	36,310	42,899	13,879	161.9%	6589	118.1%

agricultural tax that is used in Russian Federation for agricultural companies and equal to 6% of "net profit." Secondly, it was needed to determine the presence of means of production (combines, tractors, etc.) and the change dynamic for the last 3 years, as well as the number of animals employed, which is presented in Table 12.2.

Furthermore, in the analysis of the enterprise, it was necessary to review the labor resources of APC "Bolshevik." The analysis shows a total number of employees of 250 in the year of 2015, of which 210 were regular employees (tractor drivers,

Table 12.2 Number of animals employed and their cost in APC "Bolshevik"

Parameter	2013	2014	2015	2014 to 2013	2015 to 2014		2015 to 2013		
Number of animals									
Name of animals	**Units**	**Units**	**Units**	**In units**	**In %**	**In units**	**In %**	**In units**	**In %**
Cattle	3883	3865	4012	−18	99.5	147	103.8	129	103.3
Cows	1350	1350	1350	0	100.0	0	100.0	0	100.0
Milking cows	1350	1350	1350	0	100.0	0	100.0	0	100.0
Bulls	16	13	19	−3	81.3	6	146.2	3	118.8
Heifer unbred	250	258	127	8	103.2	−131	49.2	−123	50.8
Heifer 2 years and more	597	640	579	43	107.2	−61	90.5	−18	97.0
Horses	107	122	151	15	114.0	29	123.8	44	141.1
Studhorse	6	5	5	−1	83.3	0	100.0	−1	83.3
Breeding mare 3 years and more	38	38	51	0	100.0	13	134.2	13	134.2
Cost of animals (euros)									
Name of animals	**Cost**	**Cost**	**Cost**	**In RUB**	**In %**	**In RUB**	**In %**	**In RUB**	**In %**
Cattle	1,118,730	1,232,148	1,338,642	113,418	110.1	106,495	108.6	219,912	119.7
Cows	439,085	519,783	547,569	80,698	118.4	27,786	105.3	108,484	124.7
Milking cows	439,085	519,783	547,569	80,698	118.4	27,786	105.3	108,484	124.7
Bulls	8187	3791	7501	−4396	46.3	3710	197.9	−686	91.6
Heifer unbred	97,098	106,064	41,216	8966	109.2	−64,848	38.9	−55,882	42.4
Heifer 2 years and more	210,516	203,794	220,786	−6721	96.8	16,992	108.3	10,270	104.9
Horses	75,966	89,583	188,671	13,618	117.9	99,087	210.6	112,705	248.4
Studhorse	12,462	12,072	10,082	−390	96.9	−1990	83.5	−2379	80.9
Breeding mare 3 years and more	27,934	23,323	101,427	−4611	83.5	78,103	434.9	73,492	363.1
Total cost	1,194,696	1,321,731	1,527,313	281,279	110.6	293,122	115.6	574,401	127.8
Managers	34,441	35,503	35,489	1062	103.1%	−13	100.0%	1049	103.0%
Specialists	89,476	92,837	91,049	3361	103.8%	−1788	98.1%	1573	101.8%
Employees in other productions	1815	1694	1976	−121	93.3%	282	116.7%	161	108.9%

milking parlor operators, cattleman, and horse keepers), 39 were officers (managers and specialists), and 1 employee was involved in other production processes. There was no significant variation in the number of employees during the period under analysis.

The total revenues from livestock in the year 2015 amounted to 2,519,116 euros.

For a more detailed analysis of an enterprise's efficiency, it was necessary to consider such indicators as production yield per one hectare and average milk yield per one cow. In Table 12.3, the calculation of an average milk yield and yield of cereals and legumes from one hectare is presented. The data were obtained from the bookkeeping statement.

Table 12.3 Production results of specified kind of productions in APC "Bolshevik"

	Year			2014 to 2013		2015 to 2014		2015 to 2013	
Parameter	2013	2014	2015	In ha	In %	In ha	In %	In ha	In %
Total square of using lands, hectares									
Cereals total	6303	7562	8214	1186	143.9%	396	110.2%	1582	158.6%
Spring grains	2701	3887	4283	1071	142.0%	490	113.5%	1561	161.2%
Legumes grains	2551	3622	4112	115	176.7%	−94	64.5%	21	114.0%
Perennial grasses	150	265	171	0	100.0%	−228	85.0%	−228	85.0%
Annual grasses	1516	1516	1288	−311	80.0%	209	116.8%	−102	93.4%
Corn for silage and green fodder	1553	1242	1451	−533	0.0%	0	–	−533	–
Total harvest of plant production, centner (100 kg)									
Cereals total	90,168	118,050	86,302	27,882	130.9%	−31,748	73.1%	−3866	95.7%
Spring grains	87,346	112,650	84,542	25,304	129.0%	−28,108	75.0%	−2804	96.8%
Legumes grains	2822	5400	1760	2578	191.4%	−3640	32.6%	−1062	62.4%
Perennial grasses as a hay	5000	2750	5000	−2250	55.0%	2250	181.8%	0	100.0%
Perennial grasses as a green mass	217,220	124,292	80,040	−92,928	57.2%	−44,252	64.4%	−137,180	36.8%
Annual grasses	212,013	204,924	148,800	−7089	96.7%	−56,124	72.6%	−63,213	70.2%
Corn for silage and green fodder	192,020	0	0	−192,020	0.0%	0	–	−192,020	–

(continued)

Table 12.3 (continued)

	Year			2014 to 2013		2015 to 2014		2015 to 2013	
Parameter	2013	2014	2015	In ha	In %	In ha	In %	In ha	In %
Total yield of plant production, centner/hectare									
Cereals total	33.4	30,37	21,3	−3	90.9%	−9	70.1%	−12	63.8%
Spring grains	34.2	31,1	21,8	−3	90.9%	−9	70.1%	−12	63.7%
Legumes grains	18.8	20,4	10,3	2	108.5%	−10	50.5%	−9	54.8%
Perennial grasses as a hay	25.9	22	24	−4	84.9	2	109.1%	−2	92.7%
Perennial grasses as a green mass (non-applicable)	X	X	X	X	X	X	X	X	X
Annual grasses (non-applicable)	X	X	X	X	X	X	X	X	X
Corn for silage and green fodder	360.3	0	0	−360	0.0	0	–	−360	–
Total yield of milk production									
Number of milking cows, units	1350	1350	1350	0	100.0	0	100.0%	0	100.0%
Milk yield, centners	63,034	69,828	71,519	6794	110.8	1691	102.4%	8485	113.5%
Average yield of 1 cow per year, liters	4669	5172	5298	503	110.8	125	102.4%	629	113.5%
Average yield of 1 cow per day, liters	12.8	14.2	14.5	1.4	110.8	0.3	102.4%	1.7	113.5%

12.3.3 SWOT Analysis of APC “Bolshevik”

The qualitative characteristics of an enterprise help to realize production and to be economically efficient. For that, the SWOT analysis (alternatively SWOT matrix) was implemented. “A SWOT-analysis can be carried out for a company, product, place, industry, or person. It involves specifying the objective of the business venture or project and identifying the internal and external factors that are favorable and unfavorable to achieve that objective” (Osita et al. 2014).

Strengths

- High quality of management and employees.
- Presence of enough own financial resources.
- Presence of permanent markets.
- Reliable technology of goods production.
- Big opportunities for upgrading production and use of intensive production.
- Timely replacement of obsolete equipment with new one.
- Use of advanced technologies and achievements of science.
- Reliable suppliers.
- APC "Bolshevik" is the patron of the primary school in the village of Ivanovka, as well as the high school and the home of culture in the village of Spartak.

Weaknesses

- Small level of foreign economic activity.
- Underdeveloped innovative activity.
- Low quality of harvesting feed (haylage, hay).
- Inability to increase production by extensive ways.
- High electricity cost and, as a result, higher cost of milk production.
- Permanent increase of price per kilowatt.

Opportunities

- Opportunities to enter international markets, according to WTO involvement of Russia.
- Capability to implement the experience and technology know-how to new production and businesses.
- Overcoming of trading barriers on attractive foreign markets.
- Fair competition with other producers.
- Decrease of income tax on agricultural companies.
- Use of scientific achievements.
- Possibility to install renewable sources of energy. The most preferable sources are windmills and solar panels.

Threats

- Changes in climate and weather conditions, crop failure, low milk yield, and loss of cattle.
- Strengthening of positions of local competitors with equal costs.
- Emergence of new competitors with much lower costs and much higher quality of products.
- The lack of sufficient support and crop insurance from the state.
- Adverse demographic changes – lack of new employees.
- Decline of purchasing prices by dealers.
- Increase in prices of energy, fertilizers, and feeds.
- Accelerated rate of inflation.

The main threats are associated with the production of the main product – milk. The weakness is a relatively high cost of electricity, provided for agricultural companies, as well as the permanent growth of the cost of electricity. This problem is not crucial, but makes an impact on the cost of the final product – milk. High cost of energy is the most common problem for all enterprises in the world, but at the same time, quickly solvable. For the implementation of measures to increase innovative activity in APC "Bolshevik," the direction of price reduction of energy was selected, especially reduction of expenditures on electricity.

12.4 Organizational and Economic Evaluation of an Enterprise's Activity

12.4.1 Organizational Characteristic of a Portuguese Enterprise

Further consideration of practical aspects of innovative activity in an agricultural company was implemented based on the data of a Portuguese company. This company was selected from a number of agricultural enterprises as the most typical representative of its industry; according to the type of organizational structure, LDA is common for companies working in the agricultural industry in Portugal; the company belongs to a medium enterprise according to the amount of revenue, but at the same time, it's a micro-entity according to the number of employees, which is mostly common for agricultural companies in Portugal; according to the bookkeeping statements and data of the company itself, it's an innovative company, because of application of modern agricultural technologies and machineries.

Because the management of the Portuguese company has not given consent to the publication of legal data, the company will be designated as "Company B" LDA. "Company B" LDA is characterized by a linear-functional type of organizational structure, because the company is a family enterprise where only family members are involved; this itself represents the lack of any complicated organizational structure.

12.4.2 Financial and Production Characteristics of Activities of "Company B" LDA

"Company B" LDA is located in the north part of Portugal, in the subregion of Alto Trás-os-Montes, in the district of Bragança. This district is in a hard-leaved subtropical forest. The climate of the district is Mediterranean, which was formed influenced by the distance from the coast and the elevation. As the most typical representative of Mediterranean climate, in Bragança winters are relatively cold, and summers are short but very hot. Snow in winter is rare; however, in case of snowing, the snow cover can remain up for several days. The average temperature in January is 8.8 °C, and in July it is 29.2 °C. Annual precipitation, on average, is 758 mm. For the district of Bragança, water shortages during hot season and devastating forest fires in the rural areas are mostly common (IPMA 2017).

For making an estimation of economic and business efficiency of the enterprise's activity, it was necessary to analyze the following indicators: change of balance sheet structure from year to year, changes in financial results from year to year, dynamics of changes in number and amount of current assets, dynamics of changes in area and quality of lands, dynamics of changes in average number of employees, and production of various kinds of agricultural products (Horngren et al. 2012).

Firstly, the balance sheets and the income statements of "Company B" LDA were analyzed as the most important and reliable source of information. Also, both the vertical and horizontal analyses were conducted in order to get more information about the trends of the enterprise's development. The analysis of balance structure was made for 3 accounting years, 2013–2015, which were chosen because of unavailability information in the year 2016, and the data was obtained from the accounting statements.

For further analysis of business activity of the enterprise, it's necessary to analyze the income statement of "Company B" LDA for 2013–2015, as well as the vertical and horizontal analyses, to get more information about the trends of the development of the enterprise. The analysis of income statement structure was made for 3 accounting years. The data were obtained from accounting statements (*Informação Empresarial Simplificada*) and presented in Table 12.4 (year-to-year horizontal analysis). All the items in the income statement were compared with "total revenues," because in this case we can see the structure of revenues and expenditures and their influence on total income.

The next step of the analysis of business activity and state of an enterprise is to analyze its current assets. Firstly, we have analyzed the land's composition and its structure for the last 3 years, according to the data obtained from accounting statements. Secondly, we analyzed the presence of means of production (combines, tractors, etc.) and its dynamic of changing for the last 3 years, as well the number of animals employed. This information is on Table 12.5.

For further analysis of the enterprise, the labor resources of "Company B" LDA need to be looked at. This company develops its activity with only two employees.

For analysis of composition and structure of revenues from sale of production, it was also required to inspect statements with specified data from bookkeeping statements. Because the total revenues are presented only by the livestock branch, the analysis of livestock and milk production will be conducted further. The livestock revenues amount to 429,207 euros in the year 2015 (milk, 386,286 euros; livestock, 42,921 euros).

For more detailed analysis of the efficiency of the enterprise, it was necessary to consider an indicator such as average milk yield per one cow. In Table 12.6 the calculation of an average milk yield is presented.

Table 12.4 Year-to-year horizontal analysis of "Company B" LDA of income statement

Name of item	Year			2014 to 2013		2015 to 2014	
	2013	2014	2015	In amount	% of previous	In amount	% of previous
Revenues							
Plant growing	0	0	0	0,00	0.00	0,00	0.00
Livestock production	437,148	475,454	429,207	38,306	108.8%	−46,248	90.3%
Subsidies	47,697	55,702	47,699	8006	116.8%	−8004	85.6%
Revaluation of inventory	373	−954	562	−1327	−255.9%	1516	−58.9%
Works for own production	21,802	22,043	34,503	241	101.1%	12,461	156.5%
Total revenues	**507,019**	**552,245**	**511,971**	**45,227**	**108.9%**	**−40,275**	**92.7%**
Cost of production							
Plant growing	155,010	188,443	158,589	33,433	121.6%	−29,854	84.2%
Livestock production	272,014	223,005	205,231	−49,010	82.0%	−17,774	92.0%
Total cost of revenues	**427,024**	**411,448**	**363,820**	**−15,576**	**96.4%**	**−47,628**	**88.4%**
Gross profit							
Plant growing	−155,010	−188,443	−158,589	−33,433	121.6%	29,854	84.2%
Livestock production	165,133	252,449	223,976	87,316	152.9%	−28,474	88.7%
Subsidies	47,697	55,702	47,699	8006	116.8%	−8004	85.6%
Revaluation of inventory	373	−954	562	−1327	−255.9%	1516	−58.9%
Works for own production	21,802	22,043	34,503	241	101.1%	12,461	156.5%
Total gross profit	**79,995**	**140,797**	**148,150**	60,803	176.0%	7353	105.2%
Other items							
Other incomes	20,336	42,762	24,546	22,426	210.3%	−18,216	57.4%
Other expenses	9236	16,119	17,205	6883	174.5%	1086	106.7%
Profit before income taxes	**91,095**	**167,440**	**155,491**	**76,345**	**183.8%**	**−11,949**	**92.9%**
Interest income	0	0	0	0	–	0	–
Interest expense	0	2594	7322	2594	–	4728	282.2%
Other income (expense)	57,508	99,312	113,137	41,805	172.7%	13,824	113.9%
Operational profit	**33,587**	**65,534**	**35,033**	**31,947**	**195.1%**	**−30,501**	**53.5%**
Income tax	8442	10,763	6955	2321	127.5%	−3808	64.6%
Net profit	**25,145**	**54,771**	**28,078**	**29,626**	**217.8%**	**−26,693**	**51.3%**

Table 12.5 Number of animals employed and their cost in "Company B" LDA

Parameter	2013	2014	2015	2014 to 2013		2015 to 2014		2015 to 2013	
Number of animals, units									
Name of animals	**Units**	**Units**	**Units**	**In units**	**In %**	**In units**	**In %**	**In units**	**In %**
Cattle	251	241	273	−10	96.0%	32	113.3%	22	109%
Cows	112	115	164	3	102.7%	49	142.6%	52	146%
Milking cows	112	115	164	3	102.7%	49	142.6%	52	146%
Bulls	28	27	4	−1	96.4%	−23	14.8%	−24	14%
Heifer unbred	57	50	56	−7	87.7%	6	112.0%	−1	98%
Heifer 2 years and more	54	49	49	−5	90.7%	0	100.0%	−5	91%
Cost of animals (thousands of rubles – RUB)									
Name of animals	**Cost**	**Cost**	**Cost**	**In euro**	**In %**	**In euro**	**In %**	**In euro**	**In %**
Cattle	70,749	352,522	533,984	281,774	498.3%	181,461	151.5%	463,235	755%
Cows	31,569	168,216	320,781	136,647	532.8%	152,565	190.7%	289,212	1016%
Milking cows	31,569	168,216	320,781	136,647	532.8%	152,565	190.7%	289,212	1016%
Bulls	7892	39,494	7824	31,602	500.4%	−31,670	19.8%	−68	99%
Heifer unbred	16,066	73,137	109,535	57,071	455.2%	36,398	149.8%	93,469	682%
Heifer 2 years and more	15,221	71,675	95,843	56,454	470.9%	24,169	133.7%	80,622	630%
Total cost	70,749	352,522	533,984	281,774	498.3%	181,461	151.5%	463,235	755%

Table 12.6 Production results of specified kind of productions in "Company B" LDA

	Year			2014 to 2013		2015 to 2014		2015 to 2013	
Parameter	2013	2014	2015	In ha	In %	In ha	In %	In ha	In %
Total yield of milk production, liters									
Number of milking cows, units	112	115	164	3	102.7%	49	142.6%	52	146.4%
Milk yield, centners	11,395	11,822	12,710	427	103.7%	887	107.5%	1314	111.5%
Average yield of 1 cow per year, liters	10,174	10,280	7750	106	101.0%	−2530	75.4%	−2425	76.2%
Average yield of 1 cow per day, liters	27.9	28.2	21.2	0.3	101.0%	−6.9	75.4%	−6.6	76.2%

12.4.3 SWOT Analysis of "Company B" LDA

In this section, the qualitative characteristics of an enterprise are evaluated, which helps to realize production and to be economically efficient, and for that the SWOT analysis was implemented (Osita et al. 2014). The results of the implementation of this analysis are described as follows:

Strengths

- High quality of management and employees.
- Presence of enough own financial resources.
- Presence of permanent markets.
- Reliable technology of goods production.
- Timely replacement of obsolete equipment with new one.
- Reliable suppliers.

Weaknesses

- Small level of foreign economic activity.
- Underdeveloped innovative activity.
- Low quality of harvesting feed (haylage, hay).
- Inability to increase production by extensive ways.
- High electricity cost and, as a result, a growth in the cost of milk production.
- Permanent increase of price per kilowatt.

Opportunities

- Capability to implement the experience and technology know-how to new production and businesses.
- Overcoming of trading barriers on attractive foreign markets.
- Fair competition with other producers.
- Decrease of income tax on agricultural companies.
- Use of scientific achievements.
- Possibility to install renewable sources of energy. The most preferable sources are windmills and solar panels.

Threats

- Change in climate and weather conditions, crop failure, low milk yield, loss of cattle.
- Strengthening of positions of local competitors with equal costs.
- Emergence of new competitors with much lower costs and much higher quality of products.
- The lack of sufficient support and crop insurance from the state.
- Adverse demographic changes – lack of new employees.
- Decline of purchasing prices by dealers.
- Increase in prices of energy, fertilizers, and feeds.

As we can see, "Company B" LDA has approximately the same problems as APC "Bolshevik." The most problematic weakness is the high cost of electricity in Portugal. That fact leads to the opportunity to implement the system of renewable energy on the enterprise's premises, as the fastest way to reduce these kinds of expenditures.

12.5 Improving Innovative Activity in APC "Bolshevik" and "Company B" LDA

12.5.1 Recommendations for Development of Innovative Activity in APC "Bolshevik"

Based on the SWOT analysis of APC "Bolshevik" of Moskalensky district in Omsk Region, mentioned above, among the "Weaknesses" of an enterprise, such problems as high cost of electricity and its permanent growth can be observed. However, they can be solved operatively. In addition to that and based on the SWOT analysis of "Company B" LDA, among the weaknesses of the company, the high cost of electricity was also identified.

As the result of analysis of both enterprises, we can conclude that based on a common kind of activity and specialization, APC "Bolshevik" and "Company B" LDA have the point of contact in the question of conceptual directions of innovative activity development, namely, the implementation of technologies, helping to decrease the cost of electricity received from power-generating third-party companies, and the implementation of technologies, helping to completely abandon electricity produced by power-generating third-party companies and switch to the production of electricity with the capacity received as the result of implementation of these technologies.

In our opinion, the most rational way is the second option of innovative activity development in APC "Bolshevik" and in "Company B" LDA. The implementation of these technologies, by helping to produce electricity independently, will bring a positive effect in the short term, in the form of immediate decrease of electricity costs, and in the long term, in the form of additional funds that will allow the company to invest in other projects and directions of activity.

Next, the expediency of implementation of technologies will be taken into consideration, helping to produce electricity with its own capacities in APC "Bolshevik" of Moskalensky District in Omsk region. Firstly, it was necessary to track the dynamic of the cost of electricity for companies in the Omsk region from 2013 to 2015. The data were obtained from the reports of Regional Energetic Commission of Omsk region, and the results showed that the electricity cost was, in the year 2013, 0,0280/kWt and rose in the year 2015 to 0,0446/kWt (Energo-consultant 2017).

Furthermore, for estimating the necessity of implementing a new system of production and consumption of electricity, as well as to estimate the economic expediency of said system, it was necessary to conduct an analysis of consumption of electricity on total production and on specified kinds of production. For that, data from accounting statements of APC "Bolshevik" were needed, which are presented in Table 12.7 (horizontal analysis).

Because electricity is the only expense that the company can currently replace with renewable sources, it was necessary to consider the possibility of reducing its cost, which will be done by installing renewable energy sources. Based on performed SWOT analysis, opportunities for company's development

Table 12.7 Horizontal analysis of electricity cost in APC "Bolshevik"

Parameter	Year			2014 to 2013		2015 to 2014		2015 to 2013	
Type of cost	**2013**	**2014**	**2015**	**Amount**	**In %**	**Amount**	**In %**	**Amount**	**In %**
Total cost of main production	3,025,213	3,057,194	3,519,737	31,981	101%	462,543	115%	494,523	116%
In crop production	1,164,275	912,194	1,022,641	−252,081	78%	110,447	112%	−141,634	88%
In livestock production	1,860,939	2,103,475	2,413,387	242,537	113%	309,912	115%	552,449	130%
Total electricity cost	60,587	55,317	56,998	−5270	91%	1680	103%	−3589	94%
In crop production	16,360	14,653	11,252	−1707	90%	−3401	77%	−5108	69%
In livestock production	44,227	40,665	45,746	−3562	92%	5081	112%	1519	103%
Type of use of electricity	**2013**	**2014**	**2015**	**In 1000 kWt**	**In %**	**In 1000 kWt**	**In %**	**In 1000 kWt**	**In %**
Total electricity use in production:	1133	1187	1185	54	105%	−2	100%	52	105%
Total electricity use in crop	306	314	234	8	103%	−80	74%	−72	76%
Total electricity use in livestock	827	873	951	46	106%	78	109%	124	115%
Total use of electricity in production:	1005	1103	1183	98	110%	80	107%	178	118%
Total use of produced electricity	0	0	0	–	–	–	–	–	–
Total use of received electricity	1005	1103	1183	98	110%	80	107%	178	118%

were discovered – using renewable sources of energy – helping to produce electricity by company's own capacities. The most popular sources of renewable energy are windmills and solar panels. Considering that the company cannot allocate the additional squares for installing any source of energy, and based on the analysis of technical and territorial capacities of an enterprise, the following was revealed: on production squares of APC "Bolshevik," in immediate vicinity to each other, there are 14 dairy farms (barns), where, in average, there are 100 cows in each; technical characteristics of these dairy farms allow them to install solar panels on their roofs, and this installation doesn't impact their technical conditions.

Based on that, it was initially necessary to calculate the possibility of installing solar panels on the roofs of dairy farms of APC "Bolshevik." It starts with the calculation of the total surface of the roofs on which the solar panels will be installed. Based on these dimensions, it was also necessary to calculate the slope of the roof in order to find the total surface of the area. The length of the roof's side is:

$$\sqrt{\left(1^2 + 2.5^2\right)} = 2.7\,\text{m},$$

and total surface area of the roof of one dairy farm is:

$$2.7 \times 65 \times 2 = 351\,\text{m}^2.$$

Total surface of roofs of all dairy farms is:

$$351 \times 14 = 4914\,\text{m}^2.$$

On the Russian market of solar panels, there are a big number of producers. Among them there are NPP "Kvant" (Russia), "RZMKP" (Russia), "Solar Swiss International" (Switzerland), "Krasnoe Znamya – Almaz-Antey" (Russia), and many others. When choosing a supplier of solar panels, Russian producers were selected, since the products of domestic companies are not inferior to their foreign analogues in quality, yet they are relatively lower in cost (Svobodnaya energia 2017) (Invertor.ru 2017a, b). Based on the analysis of prices of solar panels in Russia and Omsk region, "Kvant KSM" of NPP "Kvant" (Russia) was selected. The main characteristics of model KSM-200 are price in euro, including VAT, 210€; capacity P_{max}, 200Wt; voltage in max power point, 36.3 V; current in max power point, 5.5A; open-circuit voltage, 45.2 V; short-circuit current, 5.9A; switching voltage, 24 V; dimensions, 1586 × 806 × 35mm; and mass, 16 kg.

To determine the required number of solar panels, it is necessary to establish their location on the roof of the barn. Because the walls of barn are made from concrete, and the roof is made from wooden racks and beams, it is possible to install the solar panels in immediate vicinity to each other. The barn construction will bear an additional load. The covered surface is:

$$2.7\,\text{m} \times 2.0\,\text{m} = 5.4\text{m}^2.$$

Based on that – for easier calculations it will be mentioned as the section – we can calculate the required number of solar panels sections:

$$4914\,\mathrm{m}^2\,/\,5.4\,\mathrm{m}^2 = 910\,\text{sections}.$$

Further, we can find the number of solar panels:

$$910\,\text{sections} \times 3\,\text{units} = 2730\,\text{units}.$$

For that, it is necessary to estimate the total cost of all panels:

$$2730 \times 210 = 573300\,\text{Euros}.$$

Also, in addition to solar panels, it is necessary to purchase batteries that will accumulate energy from the solar panels. Based on the analysis of the market of batteries for solar panels, Haze HZB batteries were chosen. Next we would like to present the models of rechargeable batteries Haze of mark HZB. The main characteristics of HZB12–230 are price in euro, including VAT, 180€; voltage, 12 V; storage capacity, 230 Ah; maximum current of charge, 57A; length, 521 mm; width, 269 mm; height, 203 mm; and mass, 70,0 kg.

Because of the biggest storage capacity, the most expedient way (in long run) was to purchase the rechargeable batteries of type Haze of mark HZB with a maximum capacity of 230 Ah. The required number of rechargeable batteries is determined by the ratio of the total capacity of solar panels to total storage capacity of rechargeable batteries as 1 to 0.375 (Invertor.ru 2017a, b). After determining the required number of solar panels, it is necessary to calculate the total amount of produced energy:

$$2730\,\text{units} \times 200\,\text{Wt} = 546000\,\text{Wt} - \text{announced capacity of all solar panels.}$$

Based on announced capacity, we can calculate the required quantity of rechargeable batteries:

$$546000\,\text{Wt}\,/\,0.375\,/\,230\,\text{Ah}\,/\,12\,\text{V} = 527\,\text{units}.$$

Further, it's necessary to determine the total cost of rechargeable batteries:

$$527 \times 210\,\text{Euros} = 94860\,\text{euros}.$$

In addition to the solar panels and rechargeable batteries, in the system of renewable energy, there are also included controllers of charge, an uninterruptible power supply unit, invertors, and other required elements. Usually, the total amount of additional goods is 1% of the total amount of solar panels and rechargeable batteries.

$$94860 + 573300 = 668160\,\text{euros}$$

$$1\% = 668160 / 100 = 6682\,\text{euros}$$

$$668160 + 6682 = 674842\,\text{euros}$$

Considering the fact that APC "Bolshevik" earned 714,987 euros net profit in 2015, and the total accumulated profit for 3 years was 1,694,018 euros, we can conclude that the enterprise has enough surplus funds for implementation and investment into renewable energy. It is necessary to notice that the management of an enterprise is trying to solve this problem.

To determine an annual amount of energy produced by these batteries, it's necessary to use the information about the average duration of daylight on the territory of Omsk region. Information will be grouped by decades – 10 days – for simplifying calculations (there will also be periods of 11 and 8 days in according months). The total approximate efficiency of using them in a standard year – 365 days – will also be calculated. According to (Dateandtime 2017), the number of hours of solar exposure in a decade is 4423,4 (hours), and the total yield in the same period is 2414 149kWt; these data need to be compared with the information from the bookkeeping statement of APC "Bolshevik" in a dynamic for 3 years. Results are in Table 12.8.

As we can see from the table, the production of electricity received from own sources (renewable energy) covers all production needs in APC "Bolshevik" with 213% in 2013 to 204% in 2015, in case we don't increase or decrease the number of solar panels. Because of the 10-year work-life of the solar panels, we can calculate the approximate results of

Table 12.8 Comparison of produced volume of electricity in dynamic of 3 years

	Year			2014 to 2013		2015 to 2014		2015 to 2013	
Indicator	2013	2014	2015	In EUR	In %	In EUR	In %	In EUR	In %
Total electricity cost, euros	60,587	55,317	56,998	−5270	91%	1680	103%	−3589	94%
Total electricity use in production, 1000 kWt	1133	1187	1185	54	105%	−2	100%	52	105%
Total electricity production by solar power, 1000 kWt	2415	2415	2415	0	100%	0	100%	0	100%
Total electricity produced by solar power, euros	129,150	112,553	116,167	−16,597	87%	3615	103%	−12,983	90%
Total cover of production needs, %	213%	203%	204%	−0,10	95%	0,00	100%	−0,09	96%

Table 12.9 Economic efficiency of renewable energy system implementation

Electricity cost	2013	2014	2015	2016	2017	2018	2019	2020	2021	2022	2023	2024	2025	2026
Cost of electricity, in euros	0.028	0.029	0.045	0.048	0.050	0.053	0.055	0.058	0.061	0.064	0.067	0.071	0.074	0.078
Annual change, in %	–	4%	54%	7%	5%	5%	5%	5%	5%	5%	5%	5%	5%	5%
Initial data, in Euros														
Total electricity cost	60,587	55,317	56,998	56,710	59,546	62,523	65,649	68,932	72,378	75,997	79,797	83,787	87,976	92,375
Total accumulated cost of electricity	–	–	–	–	59,546	122,069	187,718	256,649	329,027	405,024	484,821	568,608	656,584	748,959
Total electricity use in production, 1000 kWt	1133	1187	1185	1185	1185	1185	1185	1185	1185	1185	1185	1185	1185	1185
Implementation of the project, in Euros														
Cost of produced electricity	0	0	0	0	121,360	127,428	133,799	140,489	147,514	154,890	162,634	170,766	179,304	188,269
Accumulated cost of produced electricity	0	0	0	0	121,360	248,788	382,588	523,077	670,591	825,480	988,114	1,158,880	1,338,184	1,526,453
Production of electricity, 1000 kWt	0	0	0	0	2415	2415	2415	2415	2415	2415	2415	2415	2415	2415

(continued)

Table 12.9 (continued)

Electricity cost	2013	2014	2015	2016	2017	2018	2019	2020	2021	2022	2023	2024	2025	2026
After the project, in Euros														
Total electricity cost	60,587	55,317	56,998	56,710	−59,546	−62,523	−65,649	−68,932	−72,378	−75,997	−79,797	−83,787	−87,976	−92,375
Remained cost of electricity	0	0	0	0	61,814	64,905	68,150	71,558	75,136	78,893	82,837	86,979	91,328	95,894
Accumulated remained cost of electricity	0	0	0	0	61,814	126,719	194,870	266,428	341,563	420,456	503,293	590,272	681,600	777,495
Total non-covered electricity use in production, 1000 kWt	1133	1187	1185	1185	0	0	0	0	0	0	0	0	0	0
Remained electricity, 1000 kWt	0	0	0	0	1230	1230	1230	1230	1230	1230	1230	1230	1230	1230
Monetary terms, in Euros														
Saved funds	0	0	0	0	59,546	62,523	65,649	68,932	72,378	75,997	79,797	83,787	87,976	92,375
Accumulated saved funds	0	0	0	0	59,546	122,069	187,718	256,649	329,027	405,024	484,821	568,608	656,584	748,959
Initial expenditures and other expenses (1%)	x	x	x	−674,842	−33,742	−33,742	−33,742	−33,742	−33,742	−33,742	−33,742	−33,742	−33,742	−33,742
Final result of the project	**x**	**x**	**x**	**x**	**25,804**	**28,781**	**31,907**	**35,189**	**38,636**	**42,255**	**46,055**	**50,045**	**54,234**	**58,633**

Accumulated final result of the project	**x**	**x**	**x**	**x**	**25,804**	**54,584**	**86,491**	**121,681**	**160,317**	**202,572**	**248,627**	**298,671**	**352,905**	**411,538**
With additional income, in Euros														
Income from electricity sold, euros (25% from surplus)	0	0	0	0	15,454	16,226	17,038	17,889	18,784	19,723	20,709	21,745	22,832	23,974
Final result with additional income, euros	0	0	0	0	41,257	45,007	48,945	53,079	57,420	61,978	66,764	71,789	77,066	82,606
Accumulated new result, euros	**0**	**0**	**0**	**0**	**41,257**	**86,264**	**135,209**	**188,288**	**245,708**	**307,686**	**374,450**	**446,239**	**523,305**	**605,912**

production in kilowatts and in euro, as well as covering requirements of production. The changes in the cost of electricity were taken as average – 5% per year. The increase of 56% in 2015 was because of a very big renovation of electric grids taking place in Omsk region, and for covering these expenses, the electricity supplier company decided to increase the price for electricity. All the calculations are presented in Table 12.9.

Based on the calculation from Table 12.9, we can say that the enterprise will generate more electricity than it needs, which means that the company will generate some income from saved funds. Also, because APC "Bolshevik" will produce more electricity than it is necessary, it can sell it to Omsk region government for further resale. At the end of 2026, APC "Bolshevik" will have accumulated 411,538 euros as a result of this implementation, which the company can invest in other activities or branches. In case the company will sell all remained electricity, then the total financial result will be 605,912 euros at the end of 2026.

12.5.2 Recommendations for Development of Innovative Activity in "Company B" LDA

Next, the expediency of implementation of technologies will be considered, helping to produce electricity with its own capacities, in "Company B" LDA of Bragança Municipality of the Portuguese Republic. Firstly, it was necessary to track the dynamic of the cost of electricity for companies in Bragança Municipality from 2013 to 2015 Pordata. Data were obtained from the reports of Eurostat. The cost of electricity declined from 0,1416 euros per kWt in 2013 to 0,1402 euros per kWt in 2015. To estimate the necessity of implementing a new system of production and consumption of electricity, as well as to estimate its economic expediency, it's necessary to make the analysis of total consumption of electricity on production and on specified kinds of production. For that, it is necessary to obtain data from accounting statements of "Company B" LDA, presented in Table 12.10 (horizontal analysis).

Because electricity is the only expense that the company currently can replace by renewable sources, it is necessary to consider the possibility of reducing its cost. This will be done by installing renewable energy sources. Based on performed SWOT analysis, opportunities for company's development were discovered – using renewable sources of energy – helping to produce electricity by company's own capacities. The most popular sources of renewable energy are windmills and solar panels. Considering that the company cannot allocate additional squares for installing any source of energy, and based on the analysis of technical and territorial capacities of the enterprise, it was revealed that on production squares of "Company B" LDA, there is only one dairy farm (barn), the maximum capacity of which is 300 cows; the technical characteristics of this dairy farm allow the installation of solar panels on its roof, and this installation doesn't impact its technical conditions.

Table 12.10 Horizontal analysis of electricity cost in "Company B" LDA

Parameter	Year			2014 to 2013		2015 to 2014		2015 to 2013	
Type of cost	**2013**	**2014**	**2015**	**Amount**	**In %**	**Amount**	**In %**	**Amount**	**In %**
Total cost of main production	427,024	411,448	363,820	−15,576	96%	−47,628	88%	−63,204	85%
In crop production	155,010	188,443	158,589	33,433	122%	−29,854	84%	3579	102%
In livestock production	272,014	223,005	205,231	−49,010	82%	−17,774	92%	−66,783	75%
Total electricity cost	11,819	12,232	16,527	414	103%	4294	135%	4708	140%
In crop production	4290	5602	7204	1312	131%	1602	129%	2914	168%
In livestock production	7528	6630	9323	−899	88%	2693	141%	1794	124%
Type of use of electricity	**2013**	**2014**	**2015**	**In 1000 kWt**	**In %**	**In 1000 kWt**	**In %**	**In 1000 kWt**	**In %**
Total electricity use in production, 1000 kWt:	83	86	118	2	103%	32	138%	34	141%
Total electricity use in crop	30	39	51	9	130%	12	131%	21	170%
Total electricity use in livestock	53	46	66	−7	87%	20	143%	13	125%
Total use of electricity in production, 1000 kWt:	83	86	118	2	103%	32	138%	34	141%
Total use of produced electricity	0	0	0	–	–	–	–	–	–
Total use of received electricity	83	86	118	2	103%	32	138%	34	141%

Based on that, it is initially necessary to calculate the possibility of installing solar panels on the roof of the dairy farm of "Company B" LDA. We need to start with the calculation of the total surface on which solar panels will be installed. The slope of the roof can be calculated to find out the total area. The length of the roof's side is:

$$\sqrt{\left(1.5^2 + 10^2\right)} = 10.11\,\text{m},$$

and the total area of the roof of one dairy farm is:

$$10.11 \times 50 \times 2 = 1011\,\text{m}^2.$$

On the Portuguese market of solar panels, there is a big number of producers. Among them, there are REC. (2017) (Germany), "RZMKP" (Russia), "Solar Swiss International" (Switzerland), and many others. When making the decision about the solar panels, the prices and technical characteristics were considered (Svobodnaya energia 2017) (Invertor.ru 2017a, b).

Based on the analysis of the prices of solar panels in Bragança Municipality, REC (Germany) was selected. The electric parameters of REC solar panels are, according with standard conditions, light intensity 800 W/m^2, temperature −/+2 °C, and air mass 1.5. The main characteristics of this type of solar panel (REC196PE) are price in euros, including VAT, 175€; capacity P_{max}, 196Wt; voltage in max power point, 28.6 V; current in max power point, 6.86A; open-circuit voltage, 35.7 V; short-circuit current, 7.35A; switching voltage, 24A; dimensions, 1665 × 991 × 38mm; and mass, 18 kg.

Because of the same dimensions and weight, it is cheaper (in the long run) to buy a REC 196 PE solar panel with a maximum capacity of 196 watts. To determine the required number of solar panels, it is necessary to establish their location on the roof of the barn. Because the barn of "Company B" LDA is not made from concrete blocks and the roof is not made from wood and doesn't have enough solidity, it's impossible to install the solar panel like it is on the roofs of APC "Bolshevik." The barn is made from metal pillars with metal beams. The roof is covered by thin metal tiles. The walls and roof will bear an additional weight only if it is attributed on the points of load, with the distance between them of at least 1 meter to avoid deflection. Based on that, it is required to set the panels in the place where beams and pillars are joined together. We've considered that one solar panel will take 2 meters of the width and 1 meter of the length and named it as a section. The total area of covered surface is:

$$2.0\,\text{m} \times 1.0\,\text{m} \times 3\,\text{units} \times 17\,\text{sections} \times 2\,\text{sides} = 204\,\text{m}^2.$$

Also, we can calculate the required number of solar panels:

$$3\,\text{units} \times 17\,\text{sections} \times 2\,\text{sides} = 102\,\text{units}.$$

For that, it is necessary to estimate the total cost of all panels:

$$102\,\text{units} \times 175\,\text{euro} = 17850\,\text{euros}.$$

Also, in addition to solar panels, it is necessary to purchase batteries that will accumulate energy from the solar panels. Based on the analysis of the market of batteries for solar panels (CSB), were chosen batteries Haze HZB. Their main characteristics are price in euro, including VAT, 110€; voltage, 12 V; storage capacity, 100 Ah; maximum current of charge, 30.0A; length, 342 mm; width, 172 mm; height, 217 mm; and mass, 35.0 kg. Because of the biggest storage capacity, the most expedient way (in long-run term) was to purchase the GPL 121000 rechargeable batteries with a maximum capacity of 100 Ah. They were chosen because the grid of the solar system does not increase above 50,000 kWt, so it is not necessary to purchase batteries with big storage capacity.

The required number of rechargeable batteries is determined by the ratio of the total capacity of solar panels to total storage capacity of rechargeable batteries as 1 to 0.375 (Invertor.ru 2017a, b). After determining the required number of solar panels, it is necessary to calculate the total amount of energy produced:

$$102\,\text{units} \quad \times \quad 196\,\text{Wt} = 19\,992\,\text{Wt} - \text{announced capacity of all solar panels.}$$

Based on announced capacity we can calculate the required quantity of rechargeable batteries:

$$19\,992\,\text{Wt} / 0.375 / 100\,\text{Ah} / 12\,\text{V} = 45\,\text{units.}$$

Further, it's necessary to determine the total cost of rechargeable batteries:

$$45 \quad \times \quad 110\,\text{Euros} = 4\,950\,\text{euros.}$$

In addition to the solar panels and rechargeable batteries, in the system of renewable energy, there are also included controllers of charge, an uninterruptible power supply unit, invertors, and other required elements. Usually, the total amount of additional goods is 1% of the total amount of solar panels and rechargeable batteries.

$$17850 + 4\,950 = 22\,800\,\text{euros}$$

$$1\% = 22\,800 / 100 = 228\,\text{euros}$$

$$22\,800 + 228 = 23\,082\,\text{euros}$$

Considering the fact that "Company B" LDA earned net profit 28,078 euros in 2015 and the total accumulated profit was 107,993 euros for 3 years, we can conclude that the enterprise has enough surplus funds for implementation and investment into renewable energy. It is necessary to notice that the management of an enterprise is trying to solve this problem.

To determine an annual amount of energy produced by these batteries, it's necessary to use the information about the average duration of daylight on the territory of Bragança Municipality. Information will be grouped by decades – 10 days – for simplifying calculations (there will also be periods of 11 and 8 days in according months). The total approximate efficiency of using them in a standard year – 365 days – will also be calculated. According to (Dateandtime 2017), the number of hours of solar exposure in a decade is 4450,0 (hours), and the total yield in the same

Table 12.11 Comparison of produced volume of electricity in dynamic of 3 years

	Year			2014 to 2013		2015 to 2014		2015 to 2013	
Indicator	2013	2014	2015	Amount	In %	Amount	In %	Amount	In %
Total electricity cost, euros	11,819	12,232	16,527	414	103%	4294	135%	4708	140%
Total electricity use in production, 1000 kWt	83	86	118	2	103%	32	138%	34	141%
Total electricity production by solar power, 1000 kWt	89	89	89	0	100%	0	100%	0	100%
Total electricity production by solar power, euros	12,597	12,695	12,473	98	101%	−222	98%	−125	99%
Total cover of production needs, 1000 kWt	107%	104%	75%	−0.03	97%	−0.28	73%	−0.31	71%

period is 88 963kWt; these data need to be compared with the information from the bookkeeping statements of "Company B" LDA in a dynamic for 3 years. The results are in Table 12.11.

As we can see from the table, the production of electricity received from own sources (renewable energy) covers all use of electricity in production in "Company B" LDA, from 2013 to 2014 (from 107% to 104% accordingly), in case we don't increase the number of solar panels. But in 2015 it could not cover all needs because of a big leap of electricity consumption due to an increase in the number of animals.

Because of the 10-year work-life of the solar panels, we can calculate the approximate results of production in kilowatts and in euros, as well as covering requirements of production. The changes in the cost of electricity were taken as average – 1% per year. All the calculations are presented in Table 12.12.

Based on the calculation from Table 12.12, we can say that the enterprise will generate electricity to cover almost all needs. At the end of 2026, the accumulated funds will amount to 120,256 euros, if everything else is kept constant. The only limitation that prevents 100% coverage is the structure of the barn of "Company B" LDA. If the company strengthens the barn, it is possible to install additional solar panels on the roof to produce more electricity.

12.5.3 *Comparison of Proposed Ways to Increase Innovative Activity in APC "Bolshevik" and "Company B" LDA*

Comparison of efficiency of proposed ways to increase innovative activity between APC "Bolshevik" and "Company B" LDA needs to start from comparing the main bookkeeping items. All the data will be compared for the year 2015, as the most recent year. Data are presented in Table 12.13.

Table 12.12 Economic efficiency of renewable energy system implementation in "Company B" LDA

Parameter	2013	2014	2015	2016	2017	2018	2019	2020	2021	2022	2023	2024	2025	2026
Cost of electricity, in euros	0,142	0,143	0,140	0,140	0,142	0,143	0,144	0,146	0,147	0,149	0,150	0,152	0,153	0,155
Annual change, in %	–	1%	−2%	0%	1%	1%	1%	1%	1%	1%	1%	1%	1%	1%
Initial data, in euros														
Total electricity cost	11,819	12,232	16,527	16,527	16,692	16,859	17,027	17,198	17,370	17,543	17,719	17,896	18,075	18,256
Accumulated total cost of electricity	–	–	–	–	16,692	33,551	50,578	67,776	85,146	102,689	120,408	138,304	156,379	174,634
Total electricity use in production, 1000 kWt	83	86	118	118	118	118	118	118	118	118	118	118	118	118
Data of project, in euros														
Cost of produced electricity	0	0	0	0	12,597	12,723	12,851	12,979	13,109	13,240	13,372	13,506	13,641	13,778
Accumulated cost of produced electricity	0	0	0	0	12,597	25,321	38,171	51,150	64,259	77,499	90,872	104,378	118,019	131,797
Production of electricity, 1000 kWt	0	0	0	0	89	89	89	89	89	89	89	89	89	89
After the project, in euros														
Total electricity cost to be paid,	11,819	12,232	16,527	16,527	4095	4135	4177	4219	4261	4303	4346	4390	4434	4478
Saved funds from electricity	0	0	0		12,597	12,723	12,851	12,979	13,109	13,240	13,372	13,506	13,641	13,778
Accumulated saved funds from electricity	0	0	0	0	12,597	25,321	38,171	51,150	64,259	77,499	90,872	104,378	118,019	131,797
Total non-covered electricity use in production, 1000 kWt	83	86	118	118	29	29	29	29	29	29	29	29	29	29
Initial expenditures and other expenses (5%)	x	x	x	−23,082	−1154	−1154	−1154	−1154	−1154	−1154	−1154	−1154	−1154	−1154
Final result of the project	**x**	**x**	**x**	**x**	**11,443**	**11,569**	**11,697**	**11,825**	**11,955**	**12,086**	**12,218**	**12,352**	**12,487**	**12,623**
Accumulated final result of the project	**x**	**x**	**x**	**x**	**11,443**	**23,013**	**34,709**	**46,534**	**58,489**	**70,575**	**82,793**	**95,145**	**107,632**	**120,256**

Table 12.13 Results after implementation of projects in APC "Bolshevik" and "Company B" LDA

Item	APC "Bolshevik"	"Company B" LDA
Initial expenditures, euros	674,842	23,082
Total number of solar panels installed, units	2730	102
Capacity of all solar panels installed, kWt	546,000	19,992
Volume of electricity produced per year, 1000 kWt	2415,15	88,96
Cost of electricity produced at first year, euros	121,360	12,597
Total cover of production needs, %	204%	75%
Surplus (+)/shortage (−) of electricity for total cover, 1000 kWt	1230	−29
Saved funds per first year, euros	59,546	12,597
Saved funds in the end of project, euros	411,538	120,256
Efficiency of investments, %	61%	521%
Result in the end of the project with realization of surplus of electricity, if applicable (only 25%), euros	605,912	–
Final efficiency of investments, %	90%	521%

Based on the calculation presented in Table 12.13, we can see that after the installation of all proposed solar panels on the roofs of all barns in APC "Bolshevik," they will produce a volume of electricity two times higher than what they need. Therefore, the cover of all production needs will be 204%. That gives them the opportunity to sell the surplus to the citizens of nearby villages and generate additional income. The total efficiency of investments at the end of the project's 10-year activity (based on the work-life of solar panels) will be 90%.

At the same time, after the implementation of the project in "Company B" LDA, the company will cover through its own electricity sources with only 75% of total consumption. But still, that leads to the generating of so-called saved funds. By the end of the proposed project, the company will have saved 120,256 euros. Because of a three times higher kWt cost in Portugal compared to Russia, the efficiency of the implemented project will be almost six times higher – 521%.

All results mentioned above tell us that both proposed projects are reasonable for implementation, because they lead to generating additional and/or saved funds. That leaves it up to the company's management to make the decision to invest in the development of the company or in other activities.

12.6 Conclusions, Limitations, and Future Research Lines

Based on the implemented analysis of the definition of "innovative activity," we can conclude that it is a set of definitions that includes a lot of components related to the following branches of activity: production, storage, selling, finance, and many others. One of the main features of innovative activity is the opportunity to implement the most recent technologies, equipment, and to use the most recent information.

By conducting the analysis of activity of APC "Bolshevik," the following information was reported: company's balance sheet increased from 3,896,756 euros in 2013 to 5,235,694 euros in 2015; company's net profit increased from 373,860 euros in 2013 to 714,987 euros in 2015; company's total area of agricultural lands increased by 1911 hectares from 6303 hectares in 2013 to 8214 hectares in 2015, where 100% of it is arable; company's total number of means of production increased from 153 units in 2013 to 161 units in 2015, and in monetary terms, it increased from 1,008,512 euros in 2013 to 1,500,024 euros in 2015; the number of cattle increased by 129 units from 3833 units in 2013 to 4012 units in 2015, and in monetary terms, it increased from 83,221 euros in 2013 to 99,580 euros in 2015. The number of milking cows was the same for 2013–2015 and equal to 1350, and in monetary terms, it increased by 8070 euros from 32,663 euros in 2013 to 40,733 euros in 2015; the number of employees increased from 249 to 250 units from 2013 to 2015 accordingly, and in monetary terms, the salary increased from 2,074,815 euros in 2013 to 2,479,204 euros in 2015; total volume of milk produced increased from 57,870 centners in 2013 to 65,593 centners in 2015 and in monetary terms from 1,660,008 euros in 2013 to 2,519,116 euros in 2015.

Based on the implemented SWOT analysis, it was observed that the weaknesses include high electricity cost and the permanent growth of it.

By conducting the analysis of activity of "Company B" LDA, the following information was reported: company's balance sheet increased from 470,472 euros in 2013 to 1,039,832 euros in 2015; company's net profit increased from 25,145 euros in 2013 to 28,078 euros in 2015; company's total area of agricultural lands increased by 4,5 hectares from 17,56 hectares in 2013 to 22,06 hectares in 2015, where 100% of it is arable; company's total number of means of production remained the same, 10, but in monetary terms, it increased from 91,518 euros in 2013 to 250,148 euros in 2015; the number of cattle increased by 22 units from 251 units in 2013 to 273 units in 2015, and in monetary terms, it increased from 70,749 euros in 2013 to 533,984 euros in 2015. The number of milking cows increased from 112 units in 2013 to 164 units in 2015, and in monetary terms, it increased by 289,212 euros from 31,569 euros in 2013 to 320,781 euros in 2015; the number of employees remained the same from 2013 to 2015 and equal to two units, and in monetary terms, the salary decreased from 25,980 euros in 2013 to 25,025 euros in 2015; the total volume of milk produced and sold increased from 10,826 centners in 2013 to 12,074 centners in 2015 and in monetary terms from 393,433 euros in 2013 to 386,286 euros in 2015.

Based on the implemented SWOT analysis, it was inspected that also in the APC "Bolshevik," the weaknesses include high electricity cost and the permanent growth of it.

For solving that problem, we proposed the implementation of a new electricity-generating system, which includes changing the process of electricity production and purchasing of new required equipments. These equipments include solar panels, invertors, batteries, and additional materials. The total cost of purchasing agricultural machineries for APC "Bolshevik" is 674,842 euros and for "Company B" LDA 23082 euros. After the implementation of a new electricity-generating system, the entire electricity need for production will be covered in APC "Bolshevik" by 204% and in "Company B" LDA by 75%. That leads to reselling of surplus electricity produced by the APC "Bolshevik" to citizens of nearby villages and to the government. The total result of the implemented projects by the end of 2026 for APC "Bolshevik" is 1,526,453 euros of additional income and saved funds and for "Company B" LDA 120256 euros of saved funds.

References

Abalkin, L. (1999). *Economic encyclopedia*. Moscow: Ekonomika.

Academic. (2017). [On-line]. Available: http://dic.academic.ru/dic.nsf/ruwiki/152267#cite_note-1\, Accessed: 10.06.2017.

Azgaldov, G., & Kostin, A. (2008). Intellectual property, innovation and qualimetry. *Journal "Economic strategy"*, *2*, 162–164.

Bernar, I., Kolly, J. C., Stepanov, L., & Rybakov, V. (1997). *Explanatory glossary of economic and financial terms: French, Russian, English, German, Spanish terms* (Vol. 2). Moscow: Mezhdunarodye otnoshenia.

Bioenergy International. (2017). [On-line]. Available: http://www.infobio.ru/, Accessed: 10.06.2017.

Case Agriculture. (2017). [On-line]. Available: https://www.caseih.com/latam/pt-br, Accessed: 10.06.2017.

Choren. (2017). [On-line]. Available: http://www.choren.com/, Accessed: 10.06.2017.

CSB. (2017). [On-line]. Available: http://www.csb-battery.com/, Accessed: 10.06.2017.

Dateandtime. (2017). [On-line]. Available: http://dateandtime.info/ru/index.php, Accessed: 10.06.2017.

DeLaval. (2017). [On-line]. Available: http://www.delaval-us.com/, Accessed: 10.06.2017.

Economic and law. Encyclopedic glossary by Gabler. (1998). Moscow: Bolshaya Rossiiskaya Enciklopedia.

Energo-consultant. (2017). [On-line]. Available: http://www.energo-consultant.ru/sprav/tarifi_na_elektroenergiuy_na_2015_god/tarifi_na_elektroenergiyu_v_Omskoi_oblasti2015, Accessed: 10.06.2017.

FSSA – Federal State Statistic Authority. (2017). [On-line]. Available: http://www.gks.ru/, Accessed: 10.06.2017.

FGBE "FSCTAO named after V.I. Shumakov" of Health Ministry of Russia. (2017). [On-line]. Available: http://transpl.ru/ru/, Accessed: 10.06.2017.

Frolova, V., Kopeykin, M., & Astashova, E. (2015). Project of organizational changes in APC "Bolshevik" of Moskalensky district of Omsk region. In *Proceedings of VII International students electronic scientific conference "Students scientific forum – 2015" by RANH*. Moscow.

Garbi, E. (2002). Alternative measures of performance for e-companies: a comparison of approaches. *Journal of Business Strategies*, *19*, 1–17.

Goldstein, G. (2003). The basics of management. Study book, 2nd Edition, revised and supplemented. Taganrog: TSTU.

Google. (2017). [On-line]. Available: https://www.google.com/selfdrivingcar/, Accessed: 10.06.2017.

Gunasekaran, A., Cheng, T., & Lai, K.-H. (2006). Responsive Supply Chain: a Competitive Strategy in a Networked Economy. *Omega*, *4*, 549–564.

Horngren, C., Harrison, W., & Oliver, M. (2012). *Financial & Managerial Acoounting* (3rd ed.). Prentice Hall: Pearson Education, Inc.

Hussain, S., Ahmed, W., Rabnawaz, A., Jafar, R., Akhtar, H., & Yang, J. (2015). Supply Chain Management of Agricultural Technology Innovation: Study of Fujian and Taiwan. *Industrial Engineering Letters*, 22–30.

INE. (2017). *Instituto Nacional de Estatistica.* [On-line]. Available: https://www.ine.pt/xportal/xmain?xpgid=ine_main&xpid=INE, Accessed: 10.06.2017.

Invertor.ru. (2017a) Recommendation for selection and exploitation of solar modules. [On-line]. Available: http://invertor.ru/dop_solbat.php, Accessed: 10.06.2017.

Invertor.ru. (2017b). [On-line]. Available: http://www.invertor.ru/solbat.html, Accessed: 10.06.2017.

Investing.com – Stock market quotes and Financial news. (2017). [On-line]. Available: http://investing.com/, Accessed: 10.06.2017.

IPMA (Instituto Português do Mar e da Atmosfera). (2017). [On-line]. Available: http://www.ipma.pt/, Accessed: 10.06.2017.

Kalenskaya, N. (2012). *Innovation marketing. Study book.* Kazan: FSAEE HPE Kazan Federal University.

Lukov, V., Lukov, S., & Pogorsky, E. (2012). World practice of realization in society an innovation potential of new generations: approaches to research. *Journal "Knowledge. Understanding. Ability"*, *3*, 231–236.

NASA Earth Observatory. (2001). Precision Farming. [On-line]. Available: http://earthobservatory.nasa.gov/IOTD/view.php?id=1139, Accessed: 10.06.2017.

Osita, I. C., Onyebuchi, I. R., & Nzekwe, J. (2014). Organization's stability and productivity: the role of SWOT analysis an acronym for strength, weakness, opportunities and threat. *International Journal of Innovative and Applied Research.*, *2*, 23–32.

Plantagon. (2017). [On-line]. Available: http://www.plantagon.com/, Accessed: 10.06.2017.

Pordata. (2017). Pordata – Statistics, Charts and indicators on Municipalities, Portugal and Europe. [On-line]. Available: http://www.pordata.pt/en/Home, Accessed: 10.06.2017.

REC. (2017). [On-line]. Available: http://www.recgroup.com/en, Accessed: 10.06.2017.

Ryan, G., Bernard, H., Densin, N., & Lincoln, Y. (2000). *Data management and analysis methods in Handbook of Qualitative Research* (2nd ed.). Thousand Oaks: Sage Publication.

Schumpeter, J. (1934). *The Theory of Economic Development.* Cambridge, MA: Harvard University Press.

Schumpeter, J. (1976). Capitalism, Socialism & Democracy. 5th Edition. In *George Allen & Unwin (Publishers) Ltd.*

SpaceX. (2017). [On-line]. Available: http://spacex.com/, Accessed: 10.06.2017.

Stukach, V. (2007). *Innovative infrastructure of regional APC: study book.* Omsk: FSEE HPE OmSAU.

Svobodnaya Enegria. (2017). [On-line]. Available: http://www.solarroof.ru/, Accessed: 10.06.2017.

The Robot Report. (2017). Are ag robots ready? 27 companies profiled. [On-line]. Available: https://www.therobotreport.com/news/ag-in-transition-from-precision-ag-to-full-autonomy, Accessed: 10.06.2017.

Tseng, M.-L., Wu, K.-J., & Nguyen, T. (2011). Information technology in supply chain management: a case study. *Procedia - Social and Behavioral Sciences*, *25*, 257–272.

Zaichenko, V. (2011). Innovations: definition of meaning and civil rights of creators of innovative objects established by the legislation of Russian Federation. *Journal "Information Resources of Russia"*, *2*, 32–36.

Chapter 13
Sustainability and Innovation in the Value Chain: An Analysis of a Case Study

Ronnie J-Figueiredo, Osvaldo Luiz Gonçalves Quelhas, and Bouchaib Bahli

13.1 Introduction

Over the last decade, environmental problems related to population growth, intensification of global warming, environmental degradation, and shortage of resources are a major concern worldwide. Under the demand of economic growth – the balance between economic development, environment, and resources – they became the biggest challenge and have led many countries to seek innovative approaches to face them.

In recent decades, management practices of the green and sustainable supply chain were developed, attempting to insert environmental concerns in organizations, to reduce unintended adverse effects of production and consumption processes on the environment. In addition, the circular economy speech was spread in the literature and in the practice of industrial ecology. Circular economy extends the boundaries of environmental sustainability, by emphasizing the idea of changing products in order to build feasible relationships between ecological systems and economic growth (Genovese et al. 2017).

However, the views found in the literature emphasize the environmental pillar of sustainability and its relation with the environmental impacts generated by companies' productive processes, disregarding that an isolated approach will not be able to create a sustainable behavior for future societies. New studies should consider integration with stakeholders for shared value creation.

R. J-Figueiredo (✉) · O. L. G. Quelhas
Department of Engineering (Latec), Faculty of Engineering, Federal Fluminense University, Rio de Janeiro, Niterói, Brazil
e-mail: quelhas@latec.uff.br

B. Bahli
Department of Information Technology and Management, Faculty of Information Technology, Ryerson University, Toronto, ON, Canada
e-mail: bahli@ryerson.ca

M. Peris-Ortiz et al. (eds.), *Knowledge, Innovation and Sustainable Development in Organizations*, Innovation, Technology, and Knowledge Management,
https://doi.org/10.1007/978-3-319-74881-8_13

Besides this common approach, few studies present sustainable innovation practices for gaining competitiveness and creating shared value. Therefore, the objective of this chapter is to understand the behavior of companies in building a shared value chain for sustainable development, with a particular approach based on the economic, social, and environmental pillars of sustainability, supported by dynamic capabilities.

Dynamic capabilities can be perceived as a considerable advance in the framing and conception of internal change processes and are based on concepts such as innovation, organizational learning, and knowledge management, by promoting the stages of "reconfiguring" and "updating" the existent resources and "creating" new ones. We understand as dynamic capabilities the definition presented by Fallon-Byrne and Harney (2017), in which innovative organizations face the ongoing challenge of changing repetitive routines and behaviors for actions based on risk and uncertainty that lead to innovation.

To carry out this approach, we conducted an exploratory case study, through documentary analysis and a semi-structured interview, which allowed us to understand how an energy company's routine is connected for generating shared value.

This chapter is organized as follows: the first section presents a broad view of the concept of circular economy, the basis of sustainable development in the current global economic scenario. The second addresses innovation strategy and knowledge as key elements of regional and global competitiveness. The third explores the principles of sustainability and their elements in the context of value generation. The fourth section shows the principles of competitiveness generated by companies in the context of economic development. The fifth presents the results of the exploratory case study of an energy company, providing a view of the role of company's behavior in building a shared value chain for sustainable development and competitiveness achievement. The last section discusses the results and their future unfolding.

13.2 Circular Economy

Since the beginning of the industrial revolution, society has experienced a period of exponential economic growth that led to significant increases in living standards and organizational performance. However, the linear economic system that fueled this growth, often referred to as the "produce, consume, and dispose" model, has also brought disastrous side effects that now pose an existential threat to humanity. These range from the risk of catastrophic climate changes, exhaustion of natural resources, and economic stagnation to the destruction of natural ecosystems by pollution and unsustainable consumption (Barrie et al. 2017).

The term "linear economy," presented in the study by Kristensen et al. (Kristensen et al. 2016), was deliberately defined in order to promote its antonym, the term "circular economy." By circular, it is assumed that an economy has no net effect on the environment; instead it restores any damage made in the acquisition of resources, while ensuring that small wastes are generated throughout the production process and throughout the product's life history. The term circular has a second meaning, which relates it to the concept of cycle. There are two kinds of cycles of particular importance, the biogeochemical cycles and the idea of product recycling.

The very origin of the term "circular economy" is under debate. The idea behind a circular economy has been around for a long time. As early as 1848, Hofman, the first president of the Royal Society of Chemistry, stated: "... in an ideal chemical plant, there is, strictly speaking, no waste but only products. The better a real factory makes use of its residues, the closer it gets to its ideal, and the greater will be the profit" (Lancaster 2002).

Natural resources are indispensable for the economy and the environment to function and cooperate properly. Unfortunately, from a time when access to cheap resources seemed unlimited – a key element of economic progress over the past 200 years – it has now vanished. The widely expanded population and the increasing standard of living have created a growing demand for natural resources which we all depend on, such as metals, minerals, and food, resulting in scarcity and price increase. By the end of the next decade, two billion people in developing countries will have an income equivalent to that of the middle class and will also aspire to the lifestyle of developed countries (Wysokińska 2016).

An opportunity to create a new behavior on consumption is presented by Takahashi et al. (2017), whose results confirm the evidence already shown by other authors, that dynamic capabilities affect the organizational performance, when mediated by marketing capacity.

Teece et al. (2016) observe that distinct approaches on dynamic capabilities in the managerial literature suggest that firms should obstinately try to be agile, regardless of cost, and remain in a constant state of alert, considering that a revolutionary change occurs as a lightning, and demands continuous efforts, not just in times of crisis.

The change to the circular economy offers a perspective of sustainable growth, targeting waste and secondary raw materials, and a possibility to expand market share, going beyond increasing recycling rates and involving the entire supply chain (Vasiljevic-Shikaleska et al. 2017). Figure 13.1 shows the circular economy system diagram.

Dolinsky and Maier (2015) present a comparative relationship between circular economy and game theory, where the possibility of a circular approach is economically much more productive and cautious than a linear approach. They stress that game theory analysis is very similar to the field of sustainable development, another pillar of the relationship. They confirm the supremacy of the collaborative over the competitive way, referring to Jean Jacques Rousseau's game – the deer hunting game – where two players (hunters) can reach success if they collaborate in a situation where everybody acts individually. In short, we live in a time of constant changes, with the evolution of the 4.0 industry, 2.0 Web, and 2.0 CSR (corporate social responsibility), which demand adaptation to the individual needs of consumers.

Kristensen et al. (Kristensen et al. 2016) argue that the agrifood imaginary – eco-economy, integrated territorial agrifood, and the circular economy – have great potential for reshaping the agrifood future. The two first concepts were included as imaginary examples discussed in the literature. In the case of the circular economy, they argued that this increasingly influential approach remains

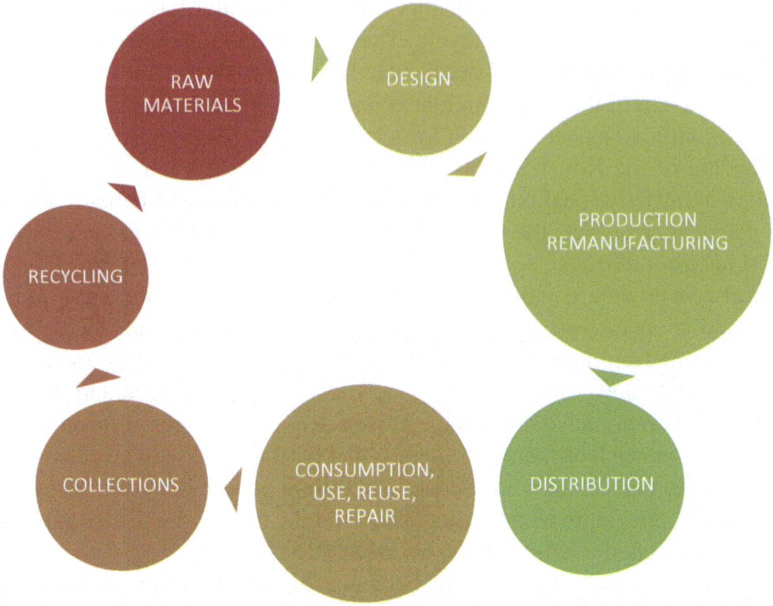

Fig. 13.1 Circular economy system diagram

unexplored in terms of its potential as food imaginary. They found that the three approaches have a common goal, but circular economy stands out due to the actors it includes, emphasizing collaborations and partnerships with existing agrifood companies.

In terms of partnership and collaboration, Djuric et al. (2017) suggest alternatives for promoting circular economy in Serbia. The results show that energy is an area that depends largely on the regional, and even global, economic, and financial situation, in addition to national policy and legislation. The less favorable the local economy, the more complex is the approach. As a sustainable strategy, they suggest to expand the country's performance in energy global markets and extend its independence by focusing primarily on the internal market.

Another case was studied by Mativenga et al. (2017), who analyzed the use of composites in a circular economy, seeking a solution for the increase in the production and volume of waste generated. They noticed that in the United Kingdom and in South Africa, there was the possibility of reducing costs, as well as the use of recycled material. Although the legislation was considered a substantial push in the United Kingdom, it was only acknowledged in SA.

Niero and Hauschild (2017) examined some common structures available for implementing circular economy strategies, and one of them is to avoid optimization of one part of the value chain (production) at the expense of other parts or to favor one category of stakeholders at the expense of others. Another issue was to identify the positive aspects of circular strategies in terms of job creation. They concluded that, given the extent of the concept of circular economy and its implications for production and consumption systems, it is crucial to go beyond the limits of different scientific disciplines in order to ensure that theory is put into practice without undesirable consequences for the economy, society, or the environment.

13.3 Innovation Strategy and Knowledge

At present, the debate on the definition of innovation strategy and its contribution to the development of economic systems lies between public and private knowledge.

Hittmar et al. (2015) define innovation strategy as a complex process that operates through the creation and implementation of new knowledge, resulting in structural organizational changes, data collection, analysis, and use of ideas to identify opportunities for innovation.

Jayaram et al. (2014) emphasize that innovation performance depends on the specific strategies and actions that the company uses to increase its innovative capacity. Specifically, in the literature of management and strategy, they found changes in leadership style, top management commitment to innovation, enhancement of people's creative skills, improvement of technological competences and breaks, and an innovative cultural orientation, which affect a company's innovation performance.

Lendel and Varmus (2011) define innovation strategy as a company's innovative approach for choosing goals, methods, and ways to use and fully develop its innovative potential. Ultimately, this approach determines the potential of innovative strategies, which can be defined as the rate of achievement of a strategy through the best use of all sources. The potential level of the innovation strategy depends on the quality of individual components and of the strategy's innovative resources. These comprise opportunities, skills, experience, invention, innovation, and especially knowledge that are available to companies in a timely manner.

In general, the most common reason for firms to access innovation strategies is the possibility of integrating knowledge workers with opportunities in a single innovation management system based on the intensive use of knowledge and economic stimulus.

According to Pavlovych and Sergiivna (2016), innovation is a relevant factor in the development of economic systems, so that economy continues to be an important task supported by the development of knowledge-intensive processes. The presence of the innovative component in production will increase the competitiveness of individual firms and of the economy as a whole, as well as the volume of production and exports. New ideas and products, advanced technologies, and organizational solutions increasingly determine the success of business activities, ensuring the survival and financial stability of the company. The need to implement innovation activities is the main requirement.

At present knowledge workers are seen as fundamental inputs for the development of new products that meet current expectations of customers and future solutions for new demands, interacting with other functions that go beyond product development, such as quality management (Kach et al. 2015). Therefore, knowledge workers are involved in complex operations and play an increasingly active role in the productive process.

The innovation strategy of a company defines the organizational priorities in the supply chain. The strategic, tactical, and operational alignment of interorganizational actions leads to innovative products, which are characterized as new and valuable and often introduced in the market. However, a managerial challenge faced by organizations is the development of supply chains capable of manufacturing products effectively, efficiently, and consistently (Jajja et al. 2017).

Although innovation is implicit in several definitions of dynamic capabilities, Alves et al. (2017) argue that innovation must be made explicit, because it is the only source of comparative advantage to sustain a firm's perpetuity. The challenge involves identifying and measuring the different forms of companies' innovation.

Felin and Powell (2016) strengthen the presented challenge with the perspective of dynamic capabilities of competitive advantage, by arguing that success in volatile industries requires superior capacities that allow companies to anticipate, shape, and adapt themselves to the new competitive scenarios. The dynamic capabilities view acknowledges the importance of resources, such as product design and manufacturing, but argues that success in these industries requires more than baseline capacities: adaptive processes and structures that allow firms to change these capacities, to anticipate changes in market demand, to develop and integrate new technologies, to learn from market events, and to foresee and capture new market opportunities.

Creating a sustainable competitive advantage through innovation requires skills, such as leadership, and appropriate supply chain practices, as reported by Ezzi and Jarboui (2016). These means questioning the context of the new knowledge economy, the selfish and opportunistic behavior of leaders, the shared structure related to diversification, the unsatisfactory strategic decisions based on shareholders' control, the lack of sufficient information to evaluate strategic decisions, the model of corporate governance and the board of directors, and developing abilities that help to build new opportunities. In short, it means to question the intangible character of the new economy and the role played by knowledge and innovation in value generation.

In contrast, Dobni et al. (2015) report that, at best, communication in organizations is difficult, and more attention should be given to forms of strategic dissemination of information about customers, competitors, and the value chain in general. Innovation ideas are guided by knowledge related to industry, value chain, competitors, and customers. It includes employees' sensitivity to the environment and the use of this improved peripheral view to identify opportunities for innovation. Many of them will be suggested by clients and suppliers and will ultimately be translated into strategic portfolio options, including horizontal and vertical integration, and comprehensive activities. Employees will also provide an internal focus on business models and processes, that is, on better ways of doing things.

One way of doing things is described by Hsu (2015), who emphasizes that, faced with the challenges of market competition, companies can effectively connect innovative research and development, marketing activities, and design resources using a value co-creation strategy (VCS) in response to innovation, marketing and design co-creation strategies, and new product development (NPD).

For Ezzi and Jarboui (2016), one of the sustainable ways to increase value by using innovation strategy is to consider the economic, environmental, and social approaches in a way that maximizes the financial performance of the business.

A connection approach is presented by Ndubisi et al. (2015), emphasizing the use of relationships between new technology ventures to improve organizational efficiency and innovation in an integrated way.

13.4 Sustainability

The fast growth of the world population together with consumption trends have led to a constant rise in the use of natural resources, increasing the pressure over the sustainability of these resources in the long range.

However, many researchers believe that this accelerated growth is unsustainable, as it leads to unintended ecological deterioration, including industrial accidents, ozone depletion, and global warming. The recent world economic crisis also accelerated the need for sustainable growth, an ecological economy, and a better use of natural resources, meeting the needs of the present without compromising the capacity of future generations (Paulraj et al. 2017).

In the 1990s, most studies addressed sustainability at the firm level, in particular, studying sustainability practices and strategies adopted by companies that had a brand license and generally coordinated activities of their supply network. Today, empirical evidence presented by Lion et al. (2016) stress the need to cross firm boundaries and assess network sustainability and its impacts on the economy, by collaborating with consumers in the value chain and with suppliers in the supply chain.

Although this need is latent, the sustainability approach is discussed by Cullen and Cullen (2017), who observe that the term is partially used in management, by emphasizing only "ecological" issues, which do not represent the whole concept – economic, social, and environmental – and is not aligned with the United Nations 17 goals of sustainable development. Figure 13.2 shows the sustainable development objectives (SDO).

This partial approach can pose a threat to the organization, as consumers are increasingly attentive to products, analyzing information on sustainability, that is, the relative environmental and social impact of the products' manufacturing and use. In many cases, consumers make decisions that involve a trade-off between

NO POVERTY	ZERO UNGER	GOOD HELATH AND WELL-BEING	QUALITY EDUCATION	GENDER EQUALITYCLEAN WATER AND SANITATION	GENDER EQUALITY
AFFORDABLE AND CLEAN ENERGY	DECENT WORK AND ECONOMIC GROWTH	INDUSTRY, INNOVATION AND INFRASTRUCTURE	REDUCED INEQUALITIES	SUSTAINABLE CITIES AND COMMUNITIES	RESPONSIBLE CONSUMPTION AND PRODUCTION
CLIMATE ACTION	LIFE BELOW WATER	LIFE ON LAND	PEACE, JUSTICE AND STRONG INSTITUTIONS	PARTNERSHIPS FOR THE GOALS	-

Fig. 13.2 Sustainable development objectives (SDO)

product sustainability and other attributes that they value. Similarly, product and marketing managers need to make decisions that reflect how consumers will respond to different trade-off scenarios, increasing the value between consumer and supplier (Luchs and Kumar 2017).

As the world becomes more aware of these consumption patterns, there are increasing pressures to reduce the impact through responsible environmental management. Public, private, and not-for-profit organizations are looking for ways to minimize their impact on the environment (Lemon, (Lemon 2017).

One of the approaches discussed by researchers is the relationship between proximity and ecological sustainability presented by Kuch (2017), who observes that this condition is never sufficient, regarding the sustainability goals. This argument stems from the observation that proximity, of all dimensions, can lead to positive impacts but also to negative ones. He contradicts a large number of researchers, by emphasizing that proximity is often interpreted as a direct source of sustainability that does not meet the objectives, since its coordination implies aligning the trajectories and state of the systems, structured and developed in an unsustainable way. He strengthens that a system developed in an unsustainable way probably will not become sustainable.

For Schilke (2014), another important issue that involves sustainability is the dynamic capabilities of companies for creating a positive learning effect through alliances. He observes that the ability to manage the alliance drives performance; however, this ability, although justified, is only partial, even after being controlled.

Although the terms sustainable, sustainability, green, and renewable are used in studies and in the market, all have a correlation with the field of energy – electrophysics, energy policies, pollution, green technologies, and energy assessment structure – where the main focus is on renewable energies, according to the results presented by Zaharia et al. (2016).

Any approximation between stakeholders should be based on guidelines that develop environmental ethics, creating progressive categories from a perspective of instrumental value – sustainable use of resources – that restricts human behavior in the natural world and from a perspective of intrinsic value, which preserves rights by putting limits on obligations related to the environment (Schuler et al. 2017).

13.5 Competitiveness

It is part of human nature to try to be successful, since the beginning. Humanity was always interested in having advantage or profit over competitors. In the course of time and based on social development, this behavior is still pronounced, and we call it competitiveness or competitive advantage.

In a traditional approach, competitiveness is considered at two levels: microeconomic (firm level) and macroeconomic (national level). At the company level, competitiveness can be interpreted as the strategic decision-making ability in the business and the rivalry between companies based on factors such as technology, innovation, skills, knowledge, human capital, etc. At the market level, national competitiveness refers to the ability to adapt policies that directly affect companies in international competition and to raise the standard of living of its citizens (Stojanovska et al. 2017).

Khan et al. (2017) show that the theory of competitive advantage rebuilds, more than a hundred years later, the theory of comparative advantage, whose emphasis was on the production of goods with comparatively low manufacturing and opportunity costs, observed in countries such as Japan, Hong Kong, and South Korea. All emphasis on competitiveness was to spread regional economic development, in light of interregional disparities in income and employment created by economic crises. Other factors like well-being, health, air quality, and safety were considered in the impact.

Regional competitiveness refers to the common features that characterize all companies in a region, including institutions, infrastructure, education, workforce skills, technology, and innovation, that is, anything that can help a company operate in a business environment conducive to its development or, on the contrary, in a hostile one. Interest in measuring regional competitiveness is due to the benefits it can bring: it provides an assessment of the local economic environment, identifies weaknesses in the local economy, encourages a long-term perspective on the economic development process, and leads to new marketing programs for the local market and the region (Clipa and Ifrim 2016).

According to Terziu (2014), the best global experience recommends new forms of regional development as a type of cluster, by concentrating similar groups in the region, since many companies have similar problems that may be related to the competitiveness of their products, to market opportunities, to the costs of sources, to quality and patterns according to market demand, etc. In short, the question is to identify opportunities for cooperation.

The most competitive regions become leaders in the global economy. They form coalitions and strong relationships with investors, strengthening their competitiveness. These regions have grown dynamically and sustainably and present high standards of living, which in turn attracts highly qualified specialists from around the world, which leads to the accumulation of significant intellectual capital. Meanwhile, regions that are less competitive struggle to become suppliers of raw materials and centers of product sales for leading regions. Regions that do not offer opportunities for development stagnate and lose their competitiveness (Zinovyeva et al. 2016). Figure 13.3 shows the 2017 IMD world competitiveness ranking.

Business sustained competitiveness differs from sustainable (Özyurt and Kantarci 2017), as it addresses sustainability as a topic present in companies since the announcement of *Our Common Future*, known as the Brundtland Report on Environment and Sustainable Development. The presented definition of sustainable development highlights a process to meet present needs without compromising future generations, which requires the cooperation of all stakeholders to improve people's quality of life, as well as the social and natural environment.

One way to understand the sustainable logic of competitiveness can be observed in the study by Aldy (2017), on the possibility of reducing carbon intensity in energy

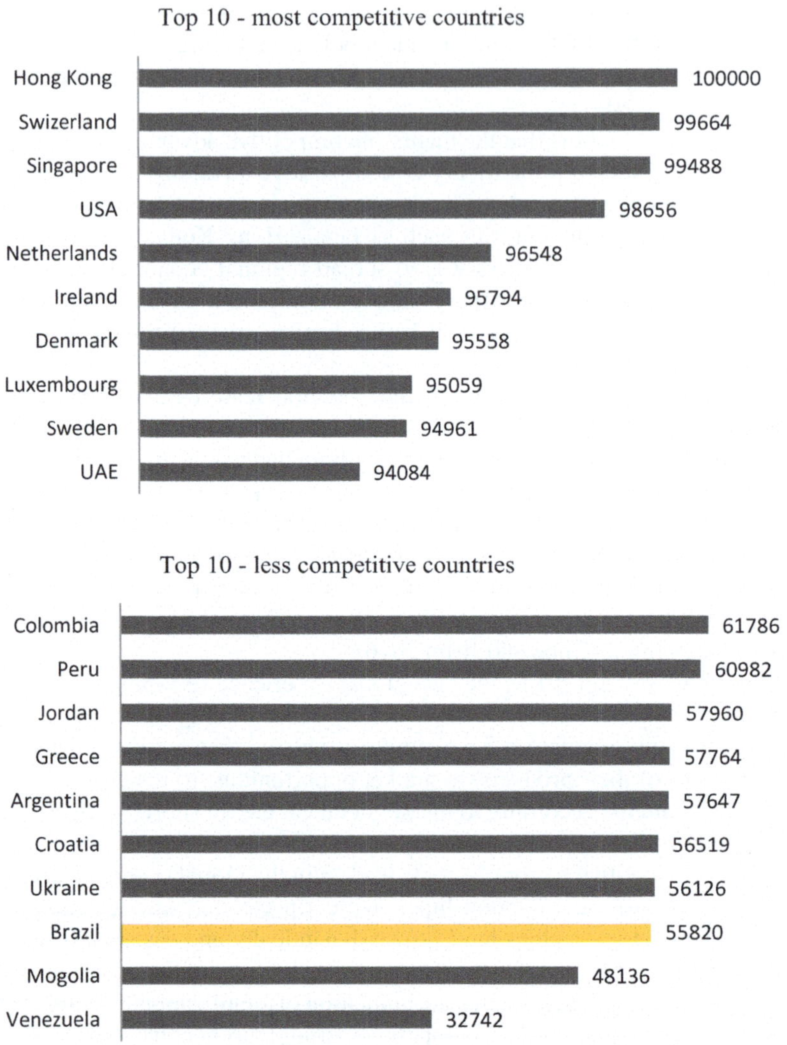

Fig. 13.3 The 2017 IMD world competitiveness ranking. (Source: IMD)

generation, changing the focus from investments in higher costs technologies, such as electricity. By defining a carbon price for fossil fuels throughout, the economy will increase energy prices. The costs of these climate policies can negatively affect domestic firms, if their competitors do not face a comparable regulation or taxation for emissions. In particular, the energy-intensive manufacturing industries expressed concern that domestic policy for climate change could impose adverse competitiveness effects, because it would raise production costs in comparison to their foreign competitors.

One of the relevant factors of regional competitiveness is the development of a global business strategy, as highlighted by Peña-Vinces et al. (2017) in their competitiveness analysis of SMEs. The research revealed that global strategy is the determining factor of competitive advantage and international competitiveness, and that it is necessary to look for competitive environments abroad, since companies gain much more confidence than if they remain in the regional environment, relying on a national strategy of differentiation and adaptation.

Another determinant factor of international competitiveness is the intensive use of knowledge, as addressed by Ochel (2002) in the study on business services, which contribute significantly to economic growth and job creation in all developed economies. They have increasingly become more international: services provided by architects and engineers, computer companies, law firms, accountants and business consultants, and advertising agents across their home country boundaries. Economists have paid close attention to the structural changes that led to a service-oriented economy, but they still largely ignore the internationalization process.

Competitiveness itself is a concept that still needs a clear definition and is still not fully understood (Rentschler and Kornejew 2017). These authors show that there is no single way of measuring competitiveness, and for this reason, they propose to evaluate the impact of energy prices on companies through an indicator defined as a "symptom" of competitiveness, which measures price variation, known as "ability to sell" or "ability to earn."

In the approach proposed by Díaz-Chao et al. (2016), the competitiveness of European regions falls into three types of dynamic agglomeration economies: first, regions as production sites, with a competitive advantage based on the availability and price of inputs; second, regions as sources of increasing returns, with a competitive advantage based on labor division and market size; and third, regions as knowledge centers, with a competitive advantage based on the quality of human resources, on the access to international markets, on the availability of business services, and on its attractiveness as a cultural and knowledge center.

13.6 Research Method

The exploratory approach made use of a case study of the company ENEL ENERGIA in order to analyze the practical situation of value generation and competitiveness gain through a shared value chain. Empirical analysis was supported by a systematic review of the literature on the topics of sustainability, circular economy, competitiveness, and strategic innovation. We applied a qualitative methodology based on analysis' tools of complex studies.

13.6.1 Research Context

As emerging countries' economies grow, they increase the use of energy and consequently expand their negative impact on the environment and society. Current problems due to the wide use of this resource cause environmental degradation and compromise future generations, mainly in terms of energy security and growth.

Increasing environmental, social, and ethical concerns, as well as a higher awareness of the effects of production and consumption on the natural environment, have led to a growing pressure from consumer organizations, environmental defense groups, and policy makers over companies, regarding social and environmental issues related to their supply chains and their products' life cycles (Zokaeï and Manikas 2014).

Since energy shortage is a threat to global energy security and, moreover, to the survival of humanity, it is critical to transform the energy sector from the traditional model into one based on renewable energy and resource saving, which is environmentally friendly and socially correct, creating a new approach for companies, based on sustainable energy practices (Zaharia et al. 2016).

ENEL ENERGIA is one of the largest private companies in the Brazilian power sector and develops renewable energy solutions. Through its activities in energy commercialization, distribution, generation, and transmission, it operates in the whole energy chain. Headquarters are located at the city of Niteroi, in the state of Rio de Janeiro, and there are other 18 operational nodes. ENEL ENERGIA has three energy distributors, located in the states of Rio de Janeiro, Ceará, and Goiás, carrying energy to around 10 million residential, commercial, industrial, rural, and public sector customers.

We highlight its leadership in the generation of solar energy in the country, both in terms of installed capacity and project portfolio. It operates the largest solar plant, in the State of Pernambuco, and is one of the largest wind energy players in the Brazilian market. It also buys and sells conventional energy in the free market, in several states, keeping a strategic asset in the conversion and transmission of energy to Mercosur (Southern Common Market).

To keep its leading role in the energy industry, it acts as an integrator of intelligent solutions, by connecting customers, homes, and companies to the technologies that are changing the traditional electrical sector – distributed generation, automation, energy storage, and electrical mobility.

The activities of conventional generation, distribution, transmission, conversion, commercialization, and energy solutions are under the control of the holding company Enel Brasil, a private limited company. In the renewable energy generation market, it operates through its subsidiary Enel Green Power Brasil (EGPB).

In order to expand our understanding of ENEL ENERGIA, we conducted a semi-structured interview with Débora Pinho, Sustainability Director.

13.7 Findings

The case description was made up from the semi-structured interview and a documentary analysis of the written material provided by the company. All information was shared publicly and free of charge in order to ensure the spreading of the business purpose, the shared value creation in the value chain.

Thus, we noticed that ENEL ENERGIA is a planned and structured company regarding the topic creating shared value (CSV). All projects involve an approximation with its stakeholders, through internal integration with the areas of business development (generation), sustainability, environment, health and safety, communication, institutional relations, engineering and construction, procurement and contracts, and mainly risk management. The company practices integration as a value in the CSV context, showing its dynamic capabilities.

In its practical approaches, the company uses a set of tools separated by concentration area: project management tools (Project Sustainability Planning and Sustainability Section in Projects Documents), analysis tools (CSV Country Analysis, Site Sustainability Checklists, Social, Economic, and Environmental Context Analysis and Stakeholders Analysis, and Social Risk Analysis), engaging tools (Stakeholder Management and Early Community Involvement), and a tool for actions design (Site Material Analysis, CSV Plan, and CSV Actions Catalogue).

We sought to understand, through a practical example, the performance of ENEL ENERGIA in the creation of shared value with the use of the value chain. Thus, we identified the case of a hydroelectric venture, which we will call the "Hidro" project.

The "Hidro" project is composed of elements that are assessed in order to test feasibility, being structured as follows: project overview, project sustainability planning, sustainability checklist, country analysis, territorial analysis, stakeholders' identification, classification, analysis and management, stakeholders' engagement, impact and risk assessment, materiality matrix, and CSV plan.

The project overview is the time of presentation of data and images related to the venture profile (existing capacity, technology, plant configuration, study stage, capital expenditures, etc.), preliminary layout of the project under study, region overview, territorial aspects (social and environmental), and delimitation of the areas of direct and indirect influence.

In planning the project sustainability, ENEL emphasizes the activities, periods, tools related to deliveries and responsible areas, according to the RACI matrix.

The checklist is based on the indicators report of the area of direct influence and contains data on geography, politics, ethnicities, development, education, health, economy, safety, public security, protected areas, and archeology, among other items.

In order to carry out country analysis, the firm uses general indicators such as area, land area, area with river, urban population, energy consumption, water quality index, capital, government, ethnic groups, religion, language, currency, and climate; economic indicators, like GDP per capita, GDP growth rate, inflation, industrial population growth rate, and electric energy consumption; social indicators, such as life expectancy at birth, unemployment rate, literacy rate, Gini coefficient, and perception of corruption index; and environmental indicators such as fresh water per capita, $CO2$ emissions per capita, land and marine protected areas, threatened species, etc.

The territorial analysis of the area of direct influence is carried out from the analysis of the social, economic, political, and environmental dimensions. Social dimension means the aspects related to the formation of the territory, social life of local communities, such as the population profile, accessibility to public service networks, to health, to education, social inclusion, as well as attributes of identity and cultural heritage, vocations, knowledge, skills, competencies, and other features of the local community that are relevant to socioeconomic development.

Economic dimension reflects the aspects related to infrastructure availability and organization of the production of goods and services, of consumption, of savings, and of investment that characterize the territory's economic activity. This dimension involves the culture of the business system, the structure of the territorial economic system, the provision of tangible and intangible infrastructure (transport networks, environmental sanitation, energy, logistics, information, and communication), and the provision of related services, technological allocation, innovation and production, and distribution processes.

Political dimension concerns the features related to the capacities and organization of public institutions to manage and affect the development process of the territories with their policies and interventions, considering the relevant aspects of the formal relationships between the institutions and the instruments used to stimulate the development of the different levels of public power, in a coordinated way.

Environmental dimension regards the aspects related to the use and protection of natural resources such as air, water, soil, and biodiversity, involving the planning of the territory for the purpose of preservation and valorization of the environmental heritage, sustainability of the development models, threats to the environment, prevention of natural disasters, and the energy issue (renewable and non-renewable resources).

The identification, classification, analysis, and management of stakeholders are carried out considering their identification (data collection campaigns, socioeconomic register, meetings, etc.), register (stakeholders' tree) and classification according to quantitative parameters of relevance (influence, dependence and tension) and qualitative (neutral, opposite and favorable). News clipping (newspapers, blogs, websites, radio, television, etc.) can be used as an important tool for the analysis of local stakeholders' statements.

Stakeholders' engagement is a dynamic and systemic way of integrating all parties during the study phase of the "Hidro" project, through the use of the following channels: meetings with field teams, information meetings, public hearings, field workers' guide, bimonthly newsletter, local radio programs, digital platform, monitoring and information center, and socioeconomic register.

The identification of the impacts related to each stage of the project is done by evaluating the level of significance of the impact (extension, duration, and frequency versus the sensitivity of the impact receive), identification of risks (description of the possible social/environmental risk generated from the selected impact and risk probability), and risk assessment (consequence of the impact and risk likelihood against the impact's level of significance).

The materiality matrix is used to identify priority themes from statements of employees and executives involved with the project, main risks of the venture, environmental impact studies, stakeholders' engagement initiatives, and clipping report.

The CSV plan, creation of shared value, is based on the following items: identification of actions and projects (project title, category, risks to be mitigated, materiality topic, and number of beneficiaries), assessment of actions and projects (benefits to the community, benefits to the company, investments estimates – costs and expenses – benefits estimates ($), potential partners, other CSV costs, and partners' funds), and action planning (responsible, media for the communication of results, partners, and suppliers, description of technical feasibility, obstacles to implementation, types of monitoring and evaluation, indicators, and time period).

Finally, it is important to notice that ENEL ENERGIA, as demonstrated through the practical "Hidro" project, has a special concern for the stakeholders, creating dynamic and competitive capabilities in the energy consumer market. Much more than simply generating financial value to shareholders, the company is committed to generate economic, environmental, and social value, in a responsible and shared way, a sustainable development for the current and mainly for future generations.

13.8 Discussion and Conclusion

Academic and business sciences suggest that sustainable development requires a greater involvement of the companies with their stakeholders, in order to allow them to adapt to the new global economic context, ensuring that decisions taken today will positively affect future generations.

One of the main approaches for companies' survival in the changing and competitive environment is to develop dynamic capabilities. To do this, the involvement of stakeholders in the shared value chain (CSV) is extremely relevant, because if they cannot guarantee the sustainability of their businesses, they will lose their image, competitiveness, and consequently market.

In recent years, there has been a significant shift in the view about the nature of the internal organizational environment and its role in the processes of strategic choice. One consequence of this change is the emergence of the resource-based view and the concept of dynamic capabilities. Although it offers attractive insights, this concept mostly remains theoretically ineffective (Krzakiewicz and Cyfert 2017).

Therefore, in our study we approached the synergy between sustainability and innovation in the value chain. To do this, we used conceptual elements such as the sustainability itself, innovation strategy, circular economy, and competitiveness, in order to understand the dynamic capabilities generated in a practical case of an electric power company.

Through this approach, we observed that dynamic capabilities were created by ENEL ENERGIA by integrating distinct actors, resources, and dimensions, thus composing the shared solution through the business value chain.

The case was conducted at a Brazilian electric power company, ENEL BRASIL S.A., a company controlled by the ENEL Group, considered one of the largest energy companies in the world and one of the five largest private companies in the Brazilian electrical sector.

The reason for choosing the company lies in its regional and international size, as well as the impact caused by the transformation of inputs into electric energy, which address current and relevant issues in the world, such as sustainability and innovation in the value chain.

The findings of the study show that ENEL ENERGIA develops, in a structured way, a planning process composed of elements that are assessed in order to configure the feasibility of hydroelectric energy projects, including the whole shared business value chain.

This planning is accomplished through the analysis of ten stages and several tools, such as project overview, project sustainability planning, sustainability checklist, country analysis, territorial analysis, identification, classification, analysis and management of stakeholders, engagement of stakeholders, impact and risk assessment, materiality matrix and CSV plan, and creating dynamic and innovative capabilities.

Our analysis of ENEL ENERGIA reveals that the company performs, in a transparent and conscious way, the integration with all interested parties, sharing knowledge, developing innovative practices, and increasing its competitiveness sustainably, keeping the principles of a circular economy, by utilizing 100 percent of the raw material used in the production process or without increasing the environmental, social, and economic impacts of the business.

The study contributed to the development of the topic "sustainability and innovation," since it evaluates, through the company's management, its dynamic capabilities to interact with the market in order to promote the sector's competitiveness and the business' sustainable development.

The exploratory approach of the case showed that the association of academic theory with the market develops solutions of high added value to businesses, noticed through the set of decisions established by ENEL ENERGIA when enabling an electric power project.

Future research could assess the costs of intangible assets involved in projects and integrate them to the report developed by the International Integrated Reporting Council to communicate the value generated by companies to society. The integrated report is consistent with developments in financial and other reports, but it shows some differences. Specifically, it focuses on an organization's ability to generate value in the short, medium, and long terms. It has a combined emphasis on conciseness, strategic focus and guidance for the future, on information connectivity, and on funds and their interdependencies. And it also stresses the importance of integrated thinking within an organization.

References

Aldy, J. E. (2017). Frameworks for evaluating policy approaches to address the competitiveness concerns of mitigating greenhouse gas emissions. *National Tax Journal, 70*(2), 395–420.

Alves, A. C., Barbieux, D., Reichert, F. M., Tello-Gamarra, J., & Zawislak, P. A. (2017). Innovation and dynamic capabilities of the firm: Defining an assessment model. *Revista de Administração de Empresas, 57*(3), 232–244 https://doi.org/10.1590/s0034-759020170304.

Barrie, J., Zawdie, G., & João, E. (2017). Leveraging triple helix and system intermediaries to enhance effectiveness of protected spaces and strategic niche management for transitioning to circular economy. *International Journal of Technology Management & Sustainable Development, 16*(1), 25–47 https://doi.org/10.1386/tmsd.16.1.25_1.

Clipa, R. I., & Ifrim, M. (2016). Measuring regional competitiveness. The case of Romania. *Annals of the University of Oradea*, 103–112.

Cullen, J. G., & Cullen, J. G. (2017). Educating business students about sustainability: A bibliometric review of current trends and research needs. 429–439. https://doi.org/10.1007/s10551-015-2838-3

Díaz-Chao, Á., Sainz-González, J., & Torrent-Sellens, J. (2016). The competitiveness of small network-firm: A practical tool. *Journal of Business Research, 69*(5), 1769–1774 https://doi.org/10.1016/j.jbusres.2015.10.053.

Djuric, S., Stosic-Mihajlovic, L., & Trajkovic, S. (2017). Circular economy and create new values: Recycling, renewable energy, ecology. *Journal of Process Management New Technologies, 5*(3), 50–68 https://doi.org/10.5937/jouproman5-14306.

Dobni, C. B., Klassen, M., & Nelson, W. T. (2015). Innovation strategy in the US: Top executives offer their views. *Journal of Business Strategy, 36*(1), 3–13 https://doi.org/10.1108/JBS-12-2013-0115.

Dolinsky, M., & Maier, S. (2015). Market-based approach in shift from linear economy towards circular economy supported by game theory analysis. *Creative and Knowledge Society, 5*(2), 1–10 https://doi.org/10.1515/cks-2015-0009.

Ezzi, F., & Jarboui, A. (2016). Does innovation strategy affect financial, social and environmental performance? *Journal of Economics Finance and Administrative Science, 21*(40), 14–24 https://doi.org/10.1016/j.jefas.2016.03.001.

Fallon-Byrne, L., & Harney, B. (2017). Microfoundations of dynamic capabilities for innovation: A review and research agenda. *The Irish Journal of Management, 36*(1), 21–32 https://doi.org/10.1515/ijm-2017-0004.

Felin, T., & Powell, T. C. (2016). Designing organizations for dynamic capabilities. *California Management Review, 58*(4), 78–96 https://doi.org/10.1525/cmr.2016.58.4.78.

Genovese, A., Acquaye, A. A., Figueroa, A., & Koh, S. C. L. (2017). Sustainable supply chain management and the transition towards a circular economy: Evidence and some applications. *Omega, 66*, 344–357 https://doi.org/10.1016/j.omega.2015.05.015.

Hittmar, S., Varmus, M., & Lendel, V. (2015). Proposal of evaluation system for successful application of innovation strategy through a set of indicators. *Procedia Economics and Finance, 26*(15), 17–22 https://doi.org/10.1016/S2212-5671(15)00796-0.

Hsu, Y. (2015). Organizational innovation strategies: The value cocreation strategy (VCS) model. *International Journal of Organizational Innovation, 8*(2), 6–20 Retrieved from http://search.ebscohost.com/login.aspx?direct=true&db=bth&AN=110553634&lang=es&site=eds-live.

Jajja, M. S. S., Kannan, V. R., Brah, S. A., & Hassan, S. Z. (2017). Linkages between firm innovation strategy, suppliers, product innovation, and business performance insights from resource dependence theory. *International Journal of Operations & Production Management, 37*(8). https://doi.org/10.1108/IJLM-11-2015-0220.

Jayaram, J., Oke, A., & Prajogo, D. (2014). The antecedents and consequences of product and process innovation strategy implementation in Australian manufacturing firms. *International Journal of Production Research, 52*(15), 4424–4439 https://doi.org/10.1080/00207543.2013.849363.

Kach, A., Azadegan, A., & Wagner, S. M. (2015). The influence of different knowledge workers on innovation strategy and product development performance in small and medium-sized enterprises. *International Journal of Production Research, 53*(8), 2489–2505. https://doi.org/Doi.https://doi.org/10.1080/00207543.2014.975856.

Khan, S., Mohammad, S., & Shah, A. (2017). Study on key empirical factors of competitiveness: Case of textile industry of Pakistan. *Journal of Managerial Sciences*, 1–23.

Kristensen, D. K., Kjeldsen, C., & ThorsÃe, M. H. (2016). Enabling sustainable agro-food futures: Exploring fault lines and synergies between the integrated territorial paradigm, rural eco-economy and circular economy. *Journal of Agricultural and Environmental Ethics, 29*(5), 749–765 https://doi.org/10.1007/s10806-016-9632-9.

Krzakiewicz, K., & Cyfert, S. (2017). Dynamic capabilities in strategic choice processes within organisations. *Dynamiczne Zdolności W Procesach Strategicznych Wyborów Organizacji., 21*(1), 7–19 https://doi.org/10.1515/manment-2015-0077.

Kuch, B. (2017). The relations between ecological sustainability and geographical proximity. *A review of the literature, 16*(1), 97–114 https://doi.org/10.1386/tmsd.16.1.97.

Lancaster, P. (2002). *Lambda-matrices and vibrating systems*. UK: Courier Corporation.
Lemon, B. M. (2017). Sustainability in the public sector. *Journal of Government Financial Management*, 18–25.
Lendel, V., & Varmus, M. (2011). Creation and implementation of the innovation strategy in the enterprise. *Economics and Management, 16*, 819–826 Retrieved from http://connection.ebscohost.com/c/articles/61822073/creation-implementation-innovation-strategy-enterprise.
Lion, A., Macchion, L., Danese, P., & Vinelli, A. (2016). Sustainability approaches within the fashion industry: The supplier perspective., *17*(2), 95–108.
Luchs, M. G., & Kumar, M. (2017). "'Yes, but this other one looks better/works better'": How do consumers respond to trade-offs between sustainability and other valued attributes? 567–584. https://doi.org/10.1007/s10551-015-2695-0
Mativenga, P. T., Sultan, A. A. M., Agwa-Ejon, J., & Mbohwa, C. (2017). Composites in a circular economy: A study of United Kingdom and South Africa. *Procedia CIRP, 61*, 691–696 https://doi.org/10.1016/j.procir.2016.11.270.
Ndubisi, N. O., Capel, C. M., & Ndubisi, G. C. (2015). Innovation strategy and performance of international technology services ventures: The moderating effect of structural autonomy. *Journal of Service Management*. https://doi.org/10.1108/MBE-09-2016-0047.
Niero, M., & Hauschild, M. Z. (2017). Closing the loop for packaging: Finding a framework to operationalize circular economy strategies. *Procedia CIRP, 61*, 685–690 https://doi.org/10.1016/j.procir.2016.11.209.
Ochel, W. (2002). The international competitiveness of business service firms: The case of Germany. *Service Industries Journal, 22*(2), 1–16 https://doi.org/10.1080/714005075.
Özyurt, P. M., & Kantarci, K. (2017). The intervening role of competitiveness on the relationship between sustainability and tourism performance: A research on European countries. *Economic Themes, 55*(1), 89–103 https://doi.org/10.1515/ethemes-2017-0006.
Paulraj, A., Chen, I. J., & Blome, C. (2017). Motives and performance outcomes of sustainable supply chain management practices: A multi-theoretical perspective. 239–258. https://doi.org/10.1007/s10551-015-2857-0
Pavlovych, D. M., & Sergiivna, D. K. (2016). Activities to enhance innovation: Innovation strategy woodworking companyEconomic Processes Management: *International Scientific E-Journal*, 4. Retrieved from http://epm.fem.sumdu.edu.ua/download/2016_4/epm2016_4_14.pdf
Peña-Vinces, J. C., Casanova, L., Guillen, J., & Urbano, D. (2017). International competitiveness of small and medium-sized enterprises: Peru, a Latin-American emerging market. *Emerging Markets Finance and Trade, 53*(1), 150–169 https://doi.org/10.1080/1540496X.2016.1156525.
Rentschler, J., & Kornejew, M. (2017). Energy price variation and competitiveness: Firm level evidence from Indonesia. *Energy Economics, 67*, 242–254 https://doi.org/10.1016/j.eneco.2017.08.015.
Schilke, O. (2014). Second-order dynamic capabilities: How do they matter? *Academy of Management Perspectives, 28*(4), 368–380 https://doi.org/10.5465/amp.2013.0093.
Schuler, D., Rasche, A., Newton, L., & Etzion, D. (2017). Corporate sustainability management and environmental ethics. *Business Ethics Quarterly, 2*(April 2017), 213–237 https://doi.org/10.1017/beq.2016.80.
Stojanovska, S., Kar, R., & Bhasin, N. (2017). The institutional determinant of competitiveness. *Journal of Sustainable Development*, *7*(17).
Takahashi, A., Bulgacov, S., Semprebon, E., & Giacomini, M. (2017). Dynamic capabilities, marketing capability and organizational performance. *Brazilian Business Review, 14*(5), 466–478 https://doi.org/10.15728/bbr.2017.14.5.1.
Teece, D., Peteraf, M., & Leih, S. (2016). Dynamic capabilities and organizational agility: Risk, uncertainty, and strategy in the innovation economy. *California Management Review, 58*(4), 13–35 https://doi.org/10.1525/cmr.2016.58.4.13.
Terziu, H. (2014). Creating clusters and competitiveness. *Iliria International Review, 1*.
Vasiljevic-Shikaleska, A., Gjozinska, B., & Stojanovikj, M. (2017). The circular economy – a pathway to sustainable future. *Journal of Sustainable Development, 7*(17), 13–30 Retrieved

from https://search.ebscohost.com/login.aspx?direct=true&db=bth&AN=123844268&site=ehost-live.
Wysokińska, Z. (2016). The "new" environmental policy of the European Union: A path to development of a circular economy and mitigation of the negative effects of climate change. *Comparative Economic Research, 19*(2), 57–73 https://doi.org/10.1515/cer-2016-0013.
Zaharia, A., Popescu, G., & Vreja, L. O. (2016). Energy scientific production in the context of the green development models. *Economic Computation and Economic Cybernetics Studies and Research, 50*(4), 151–169.
Zinovyeva, I. S., Kozenko, Y. A., Gerasimov, K. B., Dubova, Y. I., & Irizepova, M. S. (2016). Regional innovation development as a feature of competitiveness in the XXI century. *Contemporary Economics, 10*(4), 333–342 https://doi.org/10.5709/ce.1897-9254.220.
Zokaeï, K., & Manikas, I. (2014). Understanding environmental sustainability of supply chains: The biofuels industry in the United Kingdom. *Supply Chain Forum, 15*, 6–21.

Index

M. Peris-Ortiz et al. (eds.), *Knowledge, Innovation and Sustainable Development in Organizations*, Innovation, Technology, and Knowledge Management,
https://doi.org/10.1007/978-3-319-74881-8

D

E

L

M

N

O

P

Q

R

S

The manufacturer's authorised representative in the EU is Springer
Nature Customer Service Centre GmbH, Europaplatz 3, 69115 Heidelberg,
Germany. If you have any concerns regarding our products, please
contact ProductSafety@springernature.com

Printed and bound by CPI Group (UK) Ltd, Croydon, CR0 4YY
05/10/2026
02228141-0005